AF617936

Modeling and Prediction of Microgel Synthesis: Conversion, Growth and Structure

Modellierung und Prädiktion der Mikrogelsynthese: Umsatz, Wachstum und Struktur

Von der Fakultät für Maschinenwesen der Rheinisch-Westfälischen Technischen Hochschule Aachen zur Erlangung des akademischen Grades einer Doktorin der Ingenieurwissenschaften genehmigte Dissertation

vorgelegt von

Franca Alisa Laura Janssen

Berichter: Universitätsprofessor Alexander Mitsos, Ph. D.
Universitätsprofessor Dr. rer. nat. Andrij Pich

Tag der mündlichen Prüfung: 19.02.2021

Aachener Verfahrenstechnik Series – Process Systems Engineering – Volume: 15 (2021)

Franca Alisa Laura Janssen
Modeling and Prediction of Microgel Synthesis:
Conversion, Growth and Structure

Modellierung und Prädiktion der Mikrogelsynthese:
Umsatz, Wachstum und Struktur

ISBN: 978-3-95886-404-7

Bibliografische Information der Deutschen Bibliothek
Die Deutsche Bibliothek verzeichnet diese Publikation in der Deutschen Nationalbibliografie; detaillierte bibliografische Daten sind im Internet über http://dnb.ddb.de abrufbar.

Das Werk einschließlich seiner Teile ist urheberrechtlich geschützt. Jede Verwendung ist ohne die Zustimmung des Herausgebers außerhalb der engen Grenzen des Urhebergesetzes unzulässig und strafbar. Das gilt insbesondere für Vervielfältigungen, Übersetzungen, Mikroverfilmungen und die Einspeicherung und Verarbeitung in elektronischen Systemen.

Vertrieb:

1. Auflage 2021
© Verlagshaus Mainz GmbH Aachen
Süsterfeldstr. 83, 52072 Aachen
Tel. 0241/87 34 34 00
www.Verlag-Mainz.de

Herstellung:

Druckerei Mainz GmbH - Aachen
Süsterfeldstraße 83
52072 Aachen
Tel. 0241/87 34 34 00
www.DruckereiMainz.de

Satz: nach Druckvorlage des Autors
Umschlaggestaltung: Franca Janssen

printed in Germany
D 82 (Diss. RWTH Aachen University, 2021)

Vorwort

Die hier vorliegende Arbeit entstand während meiner Zeit als wissenschaftliche Mitarbeiterin der Aachener Verfahrenstechnik - Systemverfahrenstechnik der RWTH Aachen von Oktober 2013 bis 2020 und darüber hinaus.
Ich bedanke mich bei meinem Doktorvater, Herrn Prof. Alexander Mitsos, Ph. D., für die Förderung und Unterstützung in dieser Zeit. Besonders herausstellen möchte ich meinen Dank für seine guten Ideen, aber auch für viele kritische Fragen, welche mich stets herausgefordert haben, meine Arbeit zu hinterfragen.
Weiterhin danke ich Herrn Prof. Dr. rer. nat. Andrij Pich und Herrn Prof. Dr. rer. nat. Kai Leonhard sowie Michael Kather, Agnieszka Ksiazkiewicz und Leif Kröger für die stets angenehme, bereichernde und erfolgreiche Zusammenarbeit. Dieser Austausch hat meine Arbeit maßgeblich beeinflusst und in dieser Form erst möglich gemacht.
Dankbar hervorheben möchte ich auch die Zusammenarbeit mit Julian Meyer-Kirschner, Luise Kaven und Falco Jung. Durch unsere Diskussionen habe ich viel gelernt und es hat Spaß gemacht, mit Freunden zusammenzuarbeiten.
Ich möchte mich bedanken bei allen Studentinnen und Studenten, die durch HiWi-Tätigkeiten und studentische Abschlussarbeiten zu dieser Arbeit beigetragen haben. Besonderer Dank gilt dabei Johanna Kleinekorte, Daniel Felder und Steffen Linke für ihr großes Engagement.
Großer Dank geht auch an alle Mitarbeiterinnen und Mitarbeiter des Lehrstuhls. Die gute Stimmung hat mich durch diese Arbeit getragen und letztendlich dazu geführt, dass ich jeden Tag sehr gerne zur Arbeit gegangen bin. Besonders möchte ich mich für die fachliche und auch moralische Unterstützung bedanken bei Dominik Bongartz, Luisa Bree, Moll Glass, Preet Joy, Alexandra Krieger, Caroline Marks, Jennifer Puschke, Olga Walz, den Mikrogelinnen Miriam Faulde und Hanna Wolff und meinen langjährigen Bürokollegen Manuel Dahmen und Tobias Ploch. Sie alle haben meine Erinnerungen an die Zeit am Lehrstuhl deutlich geprägt.
Großer Dank geht dabei auch an Christian Redepenning für Geduld, Verständnis, Coaching, Motivation und vieles mehr.
Am meisten bedanke ich mich bei meiner Familie! Bei meinen Eltern Elisabeth und Hans-Georg, meinen Geschwistern Jana und Arne, und Mila, Nele, Elsa und Kater, die mich alle stets mit Unterstützung, Geduld und Gleichmut durch die Zeit begleitet haben.
Vielen, vielen Dank!

Köln, im Februar 2021 *Franca Janssen*

Contents

Notation XV

1 Introduction 1

2 Process model 7

2.1 Introduction and literature review . . . 7
2.2 Process model development . . . 11
2.2.1 Kinetic modeling of precipitation polymerization . . . 12
2.2.2 Energy balance of the batch reactor . . . 16
2.2.3 Experiments . . . 17
2.2.4 Quantum mechanical calculations . . . 18
2.2.5 Parameter estimation . . . 19
2.3 Results and discussion . . . 20
2.3.1 Homopolymerization . . . 20
2.3.2 Copolymerization . . . 22
2.4 Prediction of microgel properties . . . 25
2.5 Conclusion . . . 29

3 Particle size distribution model for microgel synthesis 31

3.1 Introduction and literature review . . . 31
3.2 Modeling particle size distribution for precipitation polymerization . . . 35
3.2.1 Particle nucleation and monomer partitioning . . . 37
3.2.2 Particle growth . . . 39
3.2.3 Particle coagulation . . . 40
3.2.4 Pendant double bond density . . . 41
3.2.5 Particle characterization . . . 42
3.2.6 Reaction enthalpy transfer rate . . . 43
3.3 Experimental data . . . 43
3.4 Simulation and parameter estimation . . . 44

3.5 Rcsults and discussion . . . 45
3.5.1 Simulation of reference experiment . . . 46
3.5.2 Prediction of experiment variations . . . 49
3.6 Conclusion . . . 54

4 Hybrid kinetic Monte Carlo model for internal microgel structure **57**
4.1 Introduction and literature review . . . 57
4.2 Hybrid kinetic Monte Carlo model . . . 61
4.2.1 Macro-scale . . . 63
4.2.2 Meso-scale . . . 63
4.2.3 Micro-scale . . . 64
4.2.4 Polymer structure . . . 65
4.2.5 Hybrid model algorithm . . . 67
4.3 Results and discussion . . . 70
4.3.1 Prediction of reaction enthalpy transfer rate, average microgel radius and PSD . . . 70
4.3.2 Microgel structure . . . 76
4.4 Conclusion . . . 80

5 Conclusion and future work **83**

A Parameter values **87**

Appendix **92**

B Two-phase precipitation polymerization model **93**
B.1 Liquid phase . . . 94
B.2 Gel phase . . . 95
B.3 Polymer volume . . . 96
B.4 Heat loss coefficients . . . 98
B.5 Simulation and calorimetry data for copolymerization: Direct comparison . . 98

C Case study: Precipitation terpolymerization for cross-linked PVCL-PNIPAM-based microgels **101**
C.1 Modeling of precipitation terpolymerization . . . 101
C.1.1 Quantum mechanical calculations . . . 102
C.1.2 Experiments . . . 103
C.2 Results and discussion . . . 104
C.2.1 Homo- and copolymerization of VCL and NIPAM . . . 104

C.2.2 Terpolymerization 106
C.3 Conclusion 109

D Model Equations for Pseudo-Bulk Model and additional information 111
D.1 Liquid phase modeling 111
D.2 Predicted particle radius in comparison to *in situ* DLS 113
D.3 Estimated parameter values and remarks 115
D.4 Variation of coagulation parameter values 116

E Hybrid KMC Model 119
E.1 Monomer and radical balances 119
E.2 Simulation results: Fitted hybrid KMC model 120

Bibliography 123

List of Figures

1.1 Illustration of thesis outline and approach 4

2.1 Illustration of derivation of process model from energy balance, experiments and quantum mechanical calculations 12
2.2 Schematic illustration of the two phase system in precipitation polymerization 14
2.3 Comparison of simulated and experimental enthalpy transfer rate for homopolymerization. 20
2.4 Comparison of simulated and experimental mass fractions for homopolymerization. 21
2.5 Σ_R calculated from experiments for synthesis with varying BIS concentrations 22
2.6 Simulated Σ_R for synthesis with varying cross-linker concentrations 23
2.7 Comparison of simulation with Raman spectroscopy measurements for experiments with 1.2 mol-% and 5.0 mol-% BIS 24
2.8 Simulated profiles for cross-linker, pendant double bonds and cross-links . . . 26
2.9 Contribution of gel phase as reaction locus in comparison to the fraction of gel phase in the solution 27
2.10 Cross-linking density in microgels for different initial cross-linker concentrations 28

3.1 Illustration of different particle growth mechanisms contributing to the microgel growth in precipitation polymerization 32
3.2 Comparison of experimental data and simulation with fitted model for the reference experiment 46
3.3 Predicted mean particle radius over polymerization time and measured final radius (DLS) for the reference experiment 47
3.4 Prediction of particle size distribution for the reference experiment at different stages of the process 48
3.5 Comparison of experimental and predicted reaction enthalpy transfer rates for different reaction temperatures 50

3.6 Comparison of predicted radii and measured final radii (DLS) for different reaction temperatures . . . 50
3.7 Comparison of experimental and predicted reaction enthalpy transfer rates for different initial initiator concentrations . . . 52
3.8 Comparison of predicted particle growth and measured final radii (DLS) for different initial initiator concentrations . . . 52
3.9 Comparison of experiments and predicted reaction enthalpy transfer rate for different initial cross-linker concentrations . . . 53
3.10 Comparison of predicted particle growth and measured final radii for different cross-linker concentrations . . . 53

4.1 Illustration of the three scales of the hybrid KMC model . . . 62
4.2 Scheme for the algorithm of the hybrid KMC model. . . . 68
4.3 Reaction enthalpy transfer rate calculated with hybrid model . . . 71
4.4 Average microgel radius over polymerization time, calculated with hybrid model . . . 72
4.5 Particle size distribution calculated with hybrid model for different polymerization times. . . . 73
4.6 Comparison of simulation runs with different control volumes and step sizes in terms of the reaction enthalpy transfer rate . . . 74
4.7 Comparison of simulation runs with different control volumes and step sizes in terms of the average radius . . . 75
4.8 Microgels constructed of aggregated particles . . . 77
4.9 Visualization of the final cross-link distribution for the tracked particles . . . 79

B.1 Simulated and experimental Σ_R for experiments with 1.2, 2.5 and 5.0 mol-% BIS . . . 99

C.1 Comparison of experiments and fitted model for homopolymerization of VCL in terms of Σ_R . . . 105
C.2 Comparison of experiments and fitted model for copolymerizations with BIS in terms of Σ_R . . . 105
C.3 Comparison of experiments and simulation with fitted model for terpolymerization in terms of enthalpy transfer rate . . . 107
C.4 Comparison of experiments and simulation with fitted model for terpolymerization in terms of monomer mass fractions . . . 107
C.5 Instantaneous comonomer mole fraction compared to the instantaneous normalized radius . . . 108

D.1 Prediction of growth of average particle radius over polymerization time in comparison to experimental data from *in situ* DLS 113
D.2 Simulation study for variations of coagulation parameter values r_{stable} and W to illustrate their impact on Σ_{R}, r_{h}, and PDI. 116

E.1 Reaction enthalpy transfer rate simulated with hybrid KMC model with adjusted parameter values 120
E.2 Average particle radius predicted with hybrid KMC model with adjusted parameter values 120
E.3 Particle size distribution predicted with hybrid KMC model with adjusted parameter values 121
E.4 Number of particles in the control volume V_{c} over polymerization time. . . . 122
E.5 Simulated number of particles per generation in the control volume after 500 s 122

List of Tables

2.1 Reaction mechanisms in liquid and gel phase. 13

3.1 Reaction scheme of free radical copolymerization occuring in both phases . . 36
3.2 Recipes for the reference microgel synthesis and the investigated experiment variations of different temperatures, initial initiator or cross-linker concentrations 44
3.3 Estimated parameter values for pseudo-bulk model 48

4.1 Reaction channel matrix ν for KMC algorithm 66

A.1 Parameter values for chain propagation reaction from quantum mechanical calculation 87
A.2 Parameter values for temperature dependent chain propagation reaction from quantum mechanical calculation 88
A.3 Parameter values for temperature dependent chain transfer reaction from quantum mechanical calculation 89
A.4 General parameter values or calculation thereof. 90
A.5 Estimated parameter values for copolymerization conversion model 91
A.6 Estimated parameter values from simplified copolymerization model 91

B.1 Parameter values for heat loss through reactor lid, determined from reference experiment. 98

C.1 Termination rate constants estimated with the terpolymerization model 106

D.1 Estimated parameter values and additional remarks for evaluation. 115

Notation

Variables

Symbol	Unit	Description
β		Coagulation rate
μ		Viscosity
ν		Reaction channel matrix
Σ_{R}	W	Reaction enthalpy transfer rate
τ		Event time
ϕ	%	Volume fraction
a	s^{-1}	Probability
$\mathcal{C}$		Reaction parameter
c	$\mathrm{mol\,m}^{-3}$	Concentration
F	%	Fraction
$\mathcal{F}$	$\mathrm{mol\ m}^{-}3\ \mathrm{m}^{-1}$	Density distribution
$\mathcal{H}$		Component combinations
$\mathcal{I}$		Intensity
l		Chain length
M	$\mathrm{g\,mol}^{-1}$	Molecular weight
m	g	Mass
N		Particle number
n	mol	Amount of substance
$\dot{n}$	$\mathrm{mol\,s}^{-1}$	Molar flow rate
$\bar{n}$		Average number of radicals per particle
p		Radical fraction

$\dot{Q}$	W	Heat flow
$\mathcal{R}$	$mol\,s^{-1}$	Rate
r	m	Radius
T	K	Temperature
t	s	Time
V	m^3	Volume
v	$m\,s^{-1}$	Growth rate
w		Mass fraction

Constants

Symbol	Value and Unit	Description
k_B	$1.381 \cdot 10^{-23} J\,K^{-1}$	Boltzmann constant
N_A	$6.022 \cdot 10^{23} mol^{-1}$	Avogadro constant
R	$8.314\,J\,(mol\,K)^{-1}$	Universal gas constant

Parameters

Symbol	Unit	Description
α	$W\,K^{-1}$	Heat transfer coefficient
η		Critical chain length
ρ	$kg\,m^{-3}$	Density
A	$m^3\,(mol\,s)^{-1}$	Frequency factor
c_p	$J\,(g\,K)^{-1}$	Specific heat capacity
D	$m\,s^{-1}$	Diffusion constant
E_A	$J\,mol^{-1}$	Activation energy
f		Efficiency factor
ΔH_R	$Jmol^{-1}$	Heat of reaction
k	$m^3\,(mol\,s)^{-1}$	Rate constant
P		Partition coefficient
W		Fuchs stability ratio

Math

Symbol	Description
Δ	Difference
δ	Kronecker delta function
σ	Standard deviation

Superscripts

Symbol	Description
g	Gel phase
l	Liquid phase
p	Particle

Subscripts

Symbol	Description
η	Critical chain length
λ	Moments of the radical distribution
μ	Moments of the dead polymer distribution
agg	Aggregation
avg	Average
C	Components
CM	Comonomer
c	Control
chain	Polymer chain
coag	Coagulation
d	Decomposition
des	Desorption
diff	Diffusion-limited
e	Entry
end	Simulation time end

fm	Chain transfer to monomer
h	Hydrodynamic radius
I	Initiator, Initiation
ins	Reactor inserts
inst	Instantaneous
lid	Reactor lid
loss	Heat loss
M	Monomer
max	Upper boundary
min	Lower boundary
n	Number-average
nuc	Nucleation
P	Polymer
PDB	Pendant double bond
PR	Primary radical
p	Chain propagation
part	Particle
R	Radical
reac	Reacted monomer
S	Surfactant
stable	Stable particle radius
step	Finite time step
t	Termination
td	Termination by disproportionation
tot	Total
unit	Single monomer unit in polymer chain
W	Water
w	Weight-average
X	Cross-link

Indices

Symbol	Description
Λ	Moment order
ζ, ξ	Reaction channel
$g, \tilde{g}$	Generation
i, j	Monomer species
m, n	Chain length
o	Species
q	Phase
$s, \tilde{s}$	Particle species

Acronyms

Symbol	Description
AMPA	$2,2'$-Azobis(2-methylpropionamidine) dihydrochloride
BIS	N,N'-Methylenebisacrylamide
CTAB	Cetyltrimethylammoniumbromide
DAE	Differential algebraic equations
DLS	Dynamic light scattering
DVLO theory	Derjaguin, Landau, Verwey, Overbeek theory
IHM	Indirect hard modeling
KMC	Kinetic Monte Carlo
MC	Monte Carlo
NIPAM	N-isopropylacrylamide
ODE	Ordinary differential equation
PBE	Population balance equation
PDB	Pendant double bonds
PDE	Partial differential equation
PDI	Polydispersity index
PNIPAM	Poly(N-isopropylacrylamide)

PSD	Particle size distribution
PVCL	Poly(*N*-Vinylcaprolactam)
QM	Quantum mechanical
QSSA	Quasi steady state assumption
rtc	Real-time calorimetry
TEM	Transmission electron microscopy
VCL	*N*-Vinylcaprolactam
VPTT	Volume phase transition temperature

Kurzfassung

Mikrogele sind Polymernetzwerke in der Größe von 10 - 1000 nm. In dieser Dissertation werden aufeinander aufbauend drei prädiktive Modelle für die Synthese Poly(*N*-Vinylcaprolactam)-basierter Mikrogelen durch Fällungspolymerisation entwickelt. Die drei Modelle fokussieren sich jeweils auf die wesentlichen drei Charakteristika der Mikrogelsynthese - Reaktionsfortschritt, Mikrogelwachstum und interne Struktur.

Das erste Modell ist ein vereinfachtes Zwei-Phasen System, welches der Untersuchung des Reaktionsfortschritts der Mikrogelsynthese dient. Die Integration von Parametern aus Quantenmechanischen Berechnungen sowie experimentellen Analysen bringt Einblicke in die Copolymerisation des Monomers *N*-Vinylcaprolactam mit dem Vernetzer und grundlegende Erkenntnisse über den primären Reaktionsort.

Das zweite Modell beschreibt darüber hinaus das Mikrogelwachstum sowie die erzielte Größenverteilung. Die Simulationen zeigen die Bedeutung von Radikalabsorption und Aggregation für das Mikrogelwachstum. Die in der Fällungspolymerisation typische, gleichmäßige Größenverteilung wird dabei durch eine kurze initiale Nukleationsphase erzielt, in deren Anschluss die Mikrogele gleichmäßig wachsen.

Das dritte Modell ist ein hybrides Monte Carlo Modell, welches zusätzlich auch eine Prädiktion der internen Mikrogelstruktur ermöglicht. Stochastische Simulation berechnen dabei die diskreten Aggregationen von Mikrogelen sowie jegliche Reaktionsschritte innerhalb der Mikrogele. Dieses Modell bildet die Grundlage für eine detaillierte Abbildung des Polymernetzwerks von Mikrogelen.

Die drei Modelle steigen im Detailgrad. Dabei bauen die Modelle höheren Detailgrads auf den Erkenntnissen der vorangegangenen Modelle auf. Während das erste und einfachste Modell sich besonders für die direkte Anbindung an Experimente und somit modellgestützte Prozessführung und -optimierung eignet, ist das dritte und komplexeste Modell vorrangig von prädiktiver Natur zur Ergänzung experimenteller Untersuchungen. Durch die aufeinander aufbauende Entwicklung der Modelle und Einbindungen neuer Ansätze wie Quantenmechanischer Berechnungen wird die Komplexität der Modellierung der Mikrogelsynthese gelöst.

Abstract

Microgels are cross-linked polymer networks in size of 10 - 1000 nm. In this thesis, three predictive models for the synthesis of poly(*N*-Vinylcaprolactam)-based microgels by precipitation polymerization are developed. The three models address the three characteristic properties of microgel synthesis - conversion, microgel growth and internal structure.
The first model is a reduced two-phase model, which is used for the investigation of the reaction progress and thus conversion in the microgel synthesis. The integration of parameter values from quantum mechanical calculations as well as experimental investigations elucidates the copolymerization of the monomer *N*-Vinylcaprolactam with the cross-linker. Further, it provides insight into fundamentals of the reaction, such as the primary reaction locus.
The second model describes additionally microgel growth as well as their size distribution. Simulations show the importance of radical absorption and aggregation for the microgel growth. The uniform size distribution, which is typically obtained by precipitation polymerization, is thus the result of a short initial nucleation phase, which is followed by even growth of the formed microgels.
The third model is a hybrid Kinetic Monte Carlo model. It provides additionally a prediction of the internal microgel structure. Stochastic simulation predicts the discrete aggregation of microgels as well as the individual reactions within the microgels. This model provides the fundamentals for a detailed depiction of the polymer network of microgels.
The three models increase in the degree of detail. Therein, the models of higher fidelity benefit from the knowledge gained from the previous models. While the first and simplest model is best suited for the combination with experimental analysis and thus, model-based control and optimization, the third and most complex model is rather of predictive nature to complement experimental investigations. With step-by-step model development and incorporation of new approaches such as quantum mechanical calculations, the complexity of modeling the microgel synthesis is resolved.

Chapter 1

Introduction

Microgels are smart, cross-linked polymer networks in sizes from 10 to 1000 nm. Their wide variety in shape and composition, and therefore properties and functionalities has drawn increasing attention over the past years (e.g. Saunders & Vincent (1999), Thorne et al. (2011), Plamper & Richtering (2017)). As their appearance is so diverse, they are best characterized by their common features: microgels have a stable polymer structure, are dispersed in a solvent and swollen by the solvent (Fernandez-Nieves et al. (2011)).

Microgels are "smart" materials because of their sensitivity towards changes of their environmental conditions. In their regular state, the cross-linked polymer network is swollen with solvent. The microgels are flexible and soft particles with dangling polymer chains. External stimuli, such as changes in pressure, ionic strength, pH, and temperature, cause the bonds between polymer and solvent to break, the polymer network collapses and the solvent molecules are expelled from the microgel. Within milliseconds, the microgel shrinks to a fraction of its previous size and resembles a hard colloid. As the polymer chains are cross-linked, the process of swelling and deswelling is reversible. Hence, with their form depending on their environment, microgels combine properties of hard colloids, flexible macromolecules and, - due to their interfacial properties - surfactants (Plamper & Richtering (2017)).

The morphology of microgels ranges from spherical to rod-shaped, consisting of one or multiple polymers with homogeneous to heterogeneous distribution. Even advanced structures, such as core-shell microgels with spherical layers of different polymers or Janus-microgels can be achieved. This variety in composition and shape of microgels enables a diverse range of applications. Core-shell microgels are utilized for the immobilization of nanoparticles and function as carrier systems in catalysis (Welsch et al. (2010)). Biohybrid microgels are investigated for foilar fertilizer delivery systems (Meurer et al. (2017)). Biocompatible microgels based on plasmid DNA are tested to serve as drug delivery systems

(Costa et al. (2014)). The growing interest in microgels is reflected in the number of literature reviews addressing microgels and their applications over the past years (Cortez-Lemus & Licea-Claverie (2016); Fernandez-Nieves et al. (2011); Karg et al. (2019); Plamper & Richtering (2017); Saunders et al. (2009); Thorne et al. (2011)).

Poly(*N*-vinylcaprolactam) (PVCL)-based microgels are, next to poly(*N*-isopropylacrylamide) (PNIPAM)-based microgels, the most common and therefore best investigated microgels (Cortez-Lemus & Licea-Claverie (2016)). They are swellable in water, thermo-responsive with volume phase transition temperatures (VPTT) close to the physiological temperature, and biocompatible. Further, they can easily be functionalized with diverse comonomers. These favorable properties qualify PVCL-based microgels as potential drug delivery systems for a more targeted and less invasive medical treatment (Imaz & Forcada (2010); Ramos et al. (2012)). The microgels are loaded in swollen state with guest molecules such as drugs or enzymes and function as carrier systems in the organism. The guest molecules are then released by an induced collapse or diffusion from the polymer network.

Other than biotechnological applications, PVCL-based microgels have also been investigated for coating of membranes for tunable selectivity and permeability (Lohaus et al. (2017); Menne et al. (2014)). The membrane is infiltrated with microgels in the size range of the membrane's pores. When the microgels swell due to temperature decrease below the VPTT, they block the membrane's pores and effectively decrease the membrane's permeability. Hence, the smart responsiveness of the microgels is transferred to the membrane.

For such applications, the most important properties of the microgels are size and internal cross-link distribution. The size affects the mobility of the microgel in a constricted environment, such as the penetration of pores of a membrane. The cross-link distribution determines the swelling behavior, and therefore the ability to carry and release guest molecules (Schneider et al. (2014)). Essential for any successful application of microgels is the reproducibility and adjustability of such properties. Therefore, it is required to understand the influencing factors for microgel size and internal structure, which determine the appearance and smart properties.

The microgel properties are effectively controlled by the microgel synthesis. PVCL-based microgels are usually, like PNIPAM-based microgels, synthesized in batches by (mini-)emulsion or surfactant-free precipitation polymerization (Pelton (2000); Pich & Richtering (2011)). The disperse polymer particles are formed in a heterogeneous system with a solvent-rich continuous phase. They continue to grow by polymerization as well as absorption of oligomers and coagulation with smaller microgels. The final microgels are spherical with narrow particle size distributions and statistical cross-link distribution.

Experimental studies are the common way to investigate the correlations of synthesis operation and obtained microgel properties, such as particle size and cross-link distribution (e.g. Imaz & Forcada (2010), Ramos et al. (2012), Schneider et al. (2014), Ksiazkiewicz et al. (2020)). Usually, these studies focus on the effect of individual process variables at once. However, there is a multitude of influencing factors, such as temperature, pH, solvent or feeding strategies. To investigate all influencing factors and their correlations, a high experimental effort would be required, which is intractable or at least cost and time expensive.

Mechanistic modeling can be a valuable tool to complement experimental studies. The benefits of a mechanistic model therein are diverse: Different influencing factors and their interactions can rather easily be combined and set into relation to the respective product properties in a quantitative manner. Further, a model can help to improve the understanding of the modeled process by testing hypotheses regarding the fundamental mechanisms. In combination with experimental data, unknown parameter values can be estimated, which capture the impact of influencing factors. Validated models can help reduce the experimental effort when employed for model-based design and operation of experiments.

For a mechanistic model of the microgel synthesis, three main aspects are of particular interest: (1), the overall conversion of polymer as an indicator for the progress of the synthesis; (2), the microgel growth in terms of the mean microgel size as well as their size distribution; and (3), the internal structure of the final microgels, such as the internal cross-link distribution.

However, obtaining a model that incorporates these three aspects is challenging: The microgel synthesis is a rather novel and thus a comparably unknown reaction system with a multitude of unknown parameters. The number of unknowns increases rapidly when microgels with more than just one monomer are synthesized. Polymerization typically involves numerous reactions that occur simultaneously, which again increase with the number of involved unknowns. Further, in disperse reaction systems, such as emulsion or precipitation polymerization systems, reactions may occur in two different phases. Both phases, the solvent-rich continuous phase, and the polymer-rich disperse phase, represent different reaction environments, each affecting the reaction kinetics. Finally, microgel growth might not only be the result of polymerization but also of other parallel growth mechanisms such as absorption of dissolved polymer and coagulation. While the experimental distinction between these growth mechanisms is challenging, mechanistic modeling might resolve the complexity.

In this thesis, modeling the precipitation polymerization of PVCL-based microgels is addressed. The challenges of the system complexity is overcome by sequential model development. Three different models are introduced, as illustrated in Figure 1.1. The first model focuses on the description of the polymerization reactions and hence, conversion. Therefore, the heterogeneous reaction system is described as a two-phase system with both liquid and gel phase. The second model addresses the microgel growth, including the multitude of involved growth mechanisms. The third model aims for a prediction of the internal particle structure.

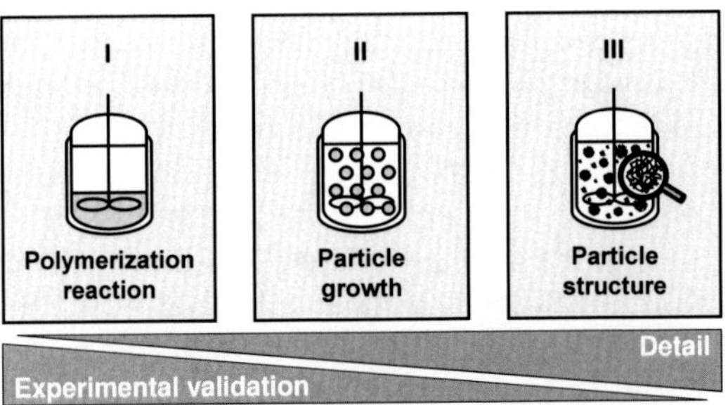

Figure 1.1: Illustration of outline and approach of this thesis: Three models with different objectives and levels of detail are formulated for precipitation polymerization of PVCL-based microgels. The level of detail increases from the first to the third model while validation from available experimental data decreases relatively.

The microgel structure of individual microgels is therein composed as the result of individual reactions and coagulation of growing particles. Where possible, the models build on their simpler representative, while increasing the level of detail. While the complexity increases from the first to the third model, the observability in terms of additionally available measurements of the predicted microgel properties and thereby the experimental validation decreases. Hence, the model with highest degree of detail predicts properties beyond experimental measurements which naturally is accompanied by a higher degree of uncertainty.

The first model for the precipitation polymerization of PVCL-based microgels is presented in Chapter 2. First, the principle assumptions regarding the polymerization mechanism are introduced. The number of unknown parameter values is reduced by employing reaction rate constants and partition coefficients from quantum mechanical calculations and transition state theory. The remaining unknown parameter values are determined by a parameter estimation with experimental data from real-time reaction calorimetry and Raman spectroscopy. The simulation results are validated with experimental data and in addition, the simulations are evaluated with regard to a coarse prediction of the radical cross-link distribution. This first model is comparably small, and therefore is best suited for parameter estimation and model-based design of experiments. Although it provides a rough explanation for the internal particle structure, it is based on rather coarse simplifications and is restricted to a prediction of the conversion.

Chapter 3 introduces the second model, which addresses microgel growth and size distribution. Based on the assumptions introduced in Chapter 2 and the parameters from quantum mechanical calculations, a pseudo-bulk model for PVCL-based microgels with cross-linker is formulated. Besides particle growth due to polymerization, the model includes the growth mechanisms radical absorption and desorption as well as coagulation. Parameter values for the different growth mechanisms are estimated based on experimental data for polymerization conversion and microgel size. With the estimated parameter values, various process operations, such as changes in temperature, cross-linker concentration and initiator concentration, are simulated and then compared to the corresponding experimental results to evaluate the predictive capabilities. This second model describes two of the key features of microgel synthesis, conversion and particle size distribution. Compared to the first model, computation time of simulation increases slightly. However, compared to the third model, it still relies on few simplifications.

In Chapter 4, the third model is presented, which targets the prediction of the microgel internal structure. Like the previously introduced models, it describes the overall polymerization progress as well as particle size distribution. In addition, it predicts the structure of the polymer network from stochastic simulation for a limited number of representative particles. The structure therein is a product of microgel growth by polymerization, radical absorption, and coagulation of smaller, growing polymer particles. In this model, the three present scales of the microgel synthesis are distinguished: (1) the reactor or macro-scale, (2) the particle or meso-scale and (3) the molecular or micro-scale. This model is the most complex and most accurate in description of growth of individual particles. However, the computation time increases due to the stochastic simulation of each reaction and aggregation event. High computation time and the embedded stochastic simulation make this model less suitable for parameter estimation in comparison to the previous models.

Each model addresses a different aspect of the microgel formation. And thus, each model has advantages and disadvantages. The first and coarsest model describes the polymerization reaction progress and while it is based on the roughest simplifications, it is best suited for a first and quick insight into the microgel formation. The finest and last model elucidates the microgel structure in great detail, but this comes at the price of high computational cost and hence, decreased usability for process optimization. Still, the development of the different models can greatly benefit from another when findings can be transferred among the models.

This thesis is, in part, based on the following publications:

- Chapter 2, which introduces the comprehensive description of the two-phase model for precipitation polymerization of PVCL-based microgels is adapted with permission from Janssen, F. A. L, Kather, M., Kröger, L. C., Mhamdi, A., Leonhard, K., Pich, A., & Mitsos, A. (2017), Synthesis of poly(*N*-vinylcaprolactam)-based microgels by precipitation polymerization: Process modeling and experimental validation. *Industrial & Engineering Chemistry Research*, **56**(49), 14545–14556. (`https://doi.org/10.`

1021/acs.iecr.7b03263). Copyright (2020) American Chemical Society.

- Chapter 3, in which the microgel growth is modeled by a pseudo-bulk model, is adapted with permission from Janssen, F. A. L., Kather, M., Ksiazkiewicz, A., Pich, A., & Mitsos, A. (2019), Synthesis of poly(*N*-vinylcaprolactam)-based microgels by precipitation polymerization: Pseudo-bulk model for particle growth and size distribution. *ACS Omega*, **4**(9), 13795–13807. (https://pubs.acs.org/doi/10.1021/acsomega.9b01335). Any further permissions related to the material excerpted is to be directed to the American Chemical Society.
- The case study in Appendix C, which conveys the approach from Chapter 2 to PVCL-PNIPAM-copolymer based microgels, is reprinted from Janssen, F. A. L., Ksiazkiewicz, A., Kather, M., Kröger, L. C., Mhamdi, A., Leonhard, K., Pich, A., & Mitsos, A. (2018), Kinetic modeling of precipitation terpolymerization for functional microgels. 109–114, of: *28th European Symposium on Computer Aided Process Engineering*. Computer Aided Chemical Engineering, vol. 43. (https://doi.org/10.1016/B978-0-444-64235-6.50021-8). With permission from Elsevier.

The manuscripts were written by Franca A. L. Janssen with recommendations for edits by the coauthors. Michael Kather and Agnieszka Ksiazkiewicz provided the experimental data used for parameter estimation. Leif C. Kröger contributed the calculations of reaction rate constants and partition coefficients. Overall, Prof. Alexander Mitsos, Ph. D. and Dr.-Ing. Adel Mhamdi provided guidance and edits to the publications and this thesis.

Chapter 2

Process model

This chapter introduces a process model for the cross-linking precipitation polymerization of PVCL-based microgels. The focus of this model is on the simulation of the polymerization reaction progress of monomer and cross-linker. The determined conversions of monomer and cross-linker are then used to provide a coarse prediction of the microgel cross-link distribution.

2.1 Introduction and literature review

Precipitation polymerization is an established method for the synthesis of PVCL-based microgels (Pich & Richtering (2011)). Polymerization of the water-soluble monomer *N*-Vinylcaprolactam (VCL) is initiated with a thermal initiator, e.g., 2,2′-Azobis(2-methylpropionamidine) dihydrochloride (AMPA). For increasing chain length of the oligomers, their solubility in water decreases. Hence, the oligomers collapse to form precursor particles. A low monomer concentration and the addition of stabilizer, e.g., Cetyltrimethylammoniumbromide (CTAB), prevent extensive aggregation of the precursor particles. At this point, the previously homogeneous system changes into a heterogeneous system with particles forming a disperse phase. Within the disperse phase, polymerization continues and particles grow. Other mechanisms such as aggregation of precursor particles and absorption of precursor particles by larger particles complement the growth of the polymer particles. The incorporation of a cross-linker, e.g., *N,N*′-Methylenebisacrylamide (BIS), leads to the formation of a stable three-dimensional polymer network. The chemical cross-links allow the gel particles to swell and collapse reversibly at volume phase transition temperature (VPTT). By exceeding the VPTT, the particle diameters of microgels, which typically range in swollen state between 50 nm and several hundred nanometers, decrease to less than half their previous size (Pich & Richtering (2011)).

The cross-linker BIS is observed to build inhomogeneously into PVCL-based microgels in batch process operation (Schneider et al. (2014)). The final particles have a highly cross-linked core and lightly cross-linked outer shell. This results in a dense particle core surrounded by dangling polymer chains (Stieger et al. (2004)). Consequently, the structure of the particles is to a large extend defined by the cross-linker distribution which can be modified by optimal operation and control of the synthesis process.

A model-based approach can support process control and operation of the synthesis, but requires reaction kinetics including the reaction rate constants. These can be determined by fitting the model to experimental data. A model for the microgel synthesis needs to address the incorporation of the cross-linker and the formation mechanism of a second phase by precipitation.

VCL and BIS react by free radical copolymerization. The copolymer composition of the resulting polymer can be calculated by the well-established *copolymer equation*, which links the composition of the polymer with the reaction kinetics and concentrations of available monomer and cross-linker (Mayo & Lewis (1944)). According to this equation, four competing propagation reactions describe the copolymerization. The reactive terminal end of the oligomer can react with either monomer or cross-linker, while the terminal end itself is either of species monomer or cross-linker. Consequently, different reaction kinetics apply. For equal monomer and cross-linker concentrations, similar reactivities between monomer and cross-linker in terms of similar reaction kinetics would result in homogeneous copolymer composition, while different reactivities result in inhomogeneous copolymer composition. Likewise, inhomogeneity can be caused by different monomer to cross-linker concentration ratios, caused by different hydrophobicities of monomer and cross-linker in the heterogeneous system.

Models for copolymerization are available with different complexities. Complex models adjust the different reaction mechanisms of free-radical polymerization to copolymerization, leading to high numbers of reactions (Dubé et al. (1997); Odian (2004)) while other approaches reformulate the copolymerization reaction to a pseudo-homopolymerization reaction (Storti et al. (1989); Xie & Hamielec (1993)).

Besides the incorporation of cross-linker by copolymerization, the formation of cross-links needs to be addressed. Cross-links are formed by propagation of a radical with a pendant double bond (PDB) of a previously incorporated cross-linker unit. By considering PDBs as an additional species, the cross-linking reactions can be calculated (Lazzari et al. (2014); Tobita & Hamielec (1989); Zhu et al. (1993)). Compared to unreacted cross-linker, the PDBs have a limited mobility, which reduces their apparent reactivity, expressed by a reduced efficiency for the cross-linking reaction. Considering the cross-linking reaction allows for the prediction of the cross-linking density, which can be employed to describe the proceeding gelation of the polymer network (Tobita & Hamielec (1989)). The calculated cross-linking density can be utilized to adjust the process operation. If the reaction kinetics are known, feeding of monomer and cross-linker in semi-batch operation can be adjusted in order to achieve a homogeneous cross-linking distribution (Enright & Zhu (2000); Hoare & McLean (2006b)). Hence, the process can be adapted to generate polymers with predefined properties.

For the investigation of the particle formation and growth mechanisms in precipitation polymerization, the reaction locus is of great interest. The experimental determination is difficult as it requires the localization of the monomer among continuous and disperse phase. Nevertheless, the modeling of a two-phase system can help understand the contribution of the two phases to the overall polymerization progress.
Different models for precipitation polymerization are proposed in literature. Some approaches simplify the precipitation mechanism, assuming that polymerization occurs mostly in the continuous phase and hence, solution polymerization kinetics can be applied (Hoare & McLean (2006a); Virtanen et al. (2015); Virtanen & Richtering (2014)). Other approaches use the contrary assumption that the continuous phase serves only as initiation phase for radicals while chain propagation is assigned only to the disperse phase (Avela et al. (1990)). As a combination of both approaches, chain length-dependent solubility of the oligomers in the continuous phase is utilized to describe the radical transfer from liquid to gel phase (Bunyakan et al. (1999)). At a discrete critical chain length, the oligomers become insoluble and precipitate. Larger oligomers are insoluble in the liquid phase, which limits the diffusion of polymer into the liquid phase, while smaller radicals partition among both phases. Although there are only few models in the literature for precipitation polymerization, the same mechanism is often applied to describe homogeneous nucleation in emulsion polymerization (e.g., Hansen & Ugelstad (1978); Immanuel et al. (2002)).

Many model-based investigations of precipitation polymerization suggest that the liquid phase functions primarily as initiation phase for the precursor particles and the disperse phase is the controlling reaction locus. One mechanism which contributes to the gel phase as preferred reaction locus is the monomer accumulation in the gel phase. As a consequence, polymer growth reactions are amplified locally in the gel phase while transfer of radicals among phases can be neglected (Arosio et al. (2011a)). On the other hand, a fundamental study of the chain length-dependent radical transfer among the phases showed that the overall mass transfer of radicals is directed from liquid to gel phase (Wieme et al. (2009)). It increases the number of radicals in the disperse phase and hence contributes to the overall polymerization. When oligomers terminate in the continuous phase before diffusing in the particle phase, both phases need to be considered as reaction loci (Mueller et al. (2005)). It needs to be considered that conclusions on the reaction locus strongly depend on the modeling assumptions concerning mass transfer among phases, e.g., partition coefficients and radical transfer. Hence, for a better prediction on the reaction locus, a comprehensive model covering the diverse transport mechanisms is favorable, but requires the individual determination of the mass transport driving mechanisms. However, these simulation studies regarding the reaction locus presume an existing disperse phase and lack a mechanism for its formation.

The combination of both mechanisms, copolymerization and precipitation, increases the number of kinetic parameters significantly. Cross-linking copolymerization increases the number of kinetic parameters by a factor three to four compared to homopolymerization due to cross-reactions among monomer and cross-linker. Modeling the two-phase system doubles the number of parameters to account for different reaction kinetics due to different surrounding conditions in the liquid and the gel phase. However, the microgel particles do not possess a rigid phase boundary but rather a fuzzy surface and even in the collapsed state, solvent remains in the microgel particles (Plamper & Richtering (2017)). This in combination with the rapid particle growth and generally low species concentrations make the experimental investigation of the two phases, e.g., measurement of the specific monomer, cross-linker and polymer concentrations, under synthesis conditions difficult.

In the literature, only few models are proposed that describe precipitation copolymerization (Arosio et al. (2011a); Costa et al. (2012); Fuxman et al. (2003)), all addressing different polymer systems. For the synthesis of microgels, only few models have been proposed, which are restricted to PNIPAM-based microgels. Hoare & McLean (2006b) proposed a multicomponent model applying solution polymerization mechanism. Based on the solution kinetics, the local distribution of functional groups and cross-links is predicted. The comparison to transmission electron microscopy (TEM) images shows that kinetics can be utilized to approximate the internal structure of the microgels (Hoare & McLean (2006a)). Though the influence of a variety of reaction parameters, including cross-linker concentration and type, on PVCL-based microgels has been studied (e.g., Imaz & Forcada (2008a); Pich et al. (2006)), a model for the synthesis of PVCL-based microgels has not been described up to this point.

In this chapter, a process model for the synthesis of PVCL-based microgels by precipitation polymerization is proposed. A standard precipitation copolymerization kinetic model is adapted to describe the free radical copolymerization of VCL and BIS with internal cross-linking. The kinetic model is complemented by an energy balance. Quantum mechanical calculations are employed to reduce the number of unknown reaction rate constants, reaction enthalpies, and partition coefficients a priori. The remaining parameter values are estimated based on reaction calorimetry measurements and Raman spectroscopy measurements of monomer and polymer mass fraction for syntheses with varying cross-linker concentrations. The determined reaction kinetics of cross-linking are used to predict the internal structure of the final microgels. This model is a first fundamental step towards the understanding of the synthesis of functional PVCL-based microgels.

2.2 Process model development

The major challenge of modeling precipitation copolymerization for PVCL-based microgels is the high number of simultaneously occurring reactions and hence, the high number of unknown reaction parameters. Unfortunately, observing the polymerization process experimentally is only possible to a limited extend. Reaction calorimetry and Raman spectroscopy are approved techniques for online measurement of the overall polymerization progress (Elizalde et al. (2005)). However, neither allows the observation of single reactions nor conclusions on the reaction locus. Hence, additional information is required, which is gained from additional experiments as well as quantum mechanical calculations.

The entire approach taken herein, from modeling to simulation, is illustrated in Figure 2.1. In the first step, a general energy balance for the reactor is formulated. Each term of the energy balance is determined experimentally, so the enthalpy transfer rate from the polymerization reaction can be extracted. A kinetic model for a two-phase copolymerization of VCL and BIS is formulated and linked to the calorimetry measurements by the reaction enthalpy and reaction rates. To describe the formation of the gel phase, the results of experiments on the chain length-dependent solubility of linear polymers are employed. Quantum mechanical calculations of the kinetic constants, partition coefficients, and reaction enthalpies reduce the number of unknown parameter values a priori.

The remaining parameter values - initiator efficiency and partition coefficient, termination rate constants, and cross-linking efficiencies -, are estimated based on reaction calorimetry and Raman spectroscopy measurements of monomer and polymer mass fraction. For the energy balance calculation, measurements of the lid temperature and the heat transfer rate to the cooling jacket are input variables to the model. The parameter estimation is performed sequentially to reduce difficulties due to correlation of the parameter values. First, the remaining parameter values for homopolymerization of VCL and BIS, respectively, are fitted to polymerization measurements of the pure compounds. The determined values are employed and the copolymerization parameter values are fitted based on experiments with varying cross-linker content.

The model is used to predict concentration profiles of the reaction compounds as well as cross-linking. Further, the growth of the disperse gel phase is simulated. The estimated cross-linking reaction kinetics are combined with the particle phase growth to approximate the internal cross-link distribution of an average microgel particle.

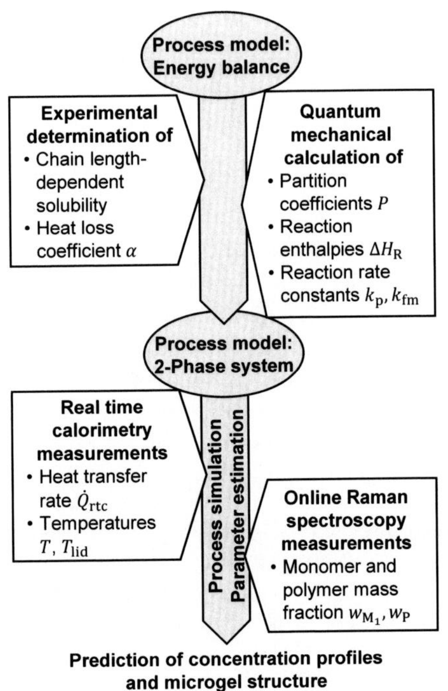

Figure 2.1: Illustration of derivation of process model from energy balance and additional information from experiments and quantum mechanical calculations.

2.2.1 Kinetic modeling of precipitation polymerization

Free radical polymerization involves various simultaneously occurring reactions. The reactions that describe precipitation polymerization of PVCL-based microgels are listed in Table 2.1. Initiator, monomer and cross-linker are dissolved in the liquid phase (l) (cf. Figure 2.2). The thermal initiator I decomposes to primary radicals $I^\bullet$, which initiate with the initiator efficiency f_I the polymerization by reaction with monomer M_j to form radicals R_1^j with a length of one repeating unit. The indices $i, j = 1, 2$ denote the type of terminal end, monomer VCL or cross-linker BIS, respectively. An oligomer R_n^i with n repeating units continues to grow by propagation with monomer and cross-linker. Further, radicals can be trans-

Table 2.1: Reaction mechanisms with associated reaction rate constants k in liquid phase (l) and gel phase (g). The indices $i, j = 1$ represent VCL, $i, j = 2$ represent BIS. The model is adapted from common polymerization models (e.g. Dubé et al. (1997); Odian (2004)) to VCL and BIS.

Liquid phase l			
Decomposition	I	$\xrightarrow{k_\mathrm{d}^\mathrm{l}}$	$2\mathrm{I}^\bullet$
Initiation	$\mathrm{I}^\bullet + \mathrm{M}_j$	$\xrightarrow{k_\mathrm{I}^\mathrm{l}}$	$\mathrm{R}_1^j + (j-1)\mathrm{PDB}$
Propagation	$\mathrm{R}_n^i + \mathrm{M}_j$	$\xrightarrow{k_{\mathrm{p}ij}^\mathrm{l}}$	$\mathrm{R}_{n+1}^j + (j-1)\mathrm{PDB}$
Chain transfer to monomer	$\mathrm{R}_n^i + \mathrm{M}_j$	$\xrightarrow{k_{\mathrm{fm}ij}^\mathrm{l}}$	$\mathrm{P}_n + \mathrm{R}_1^j + (j-1)\mathrm{PDB}$
Termination by disproportionation	$\mathrm{R}_n^i + \mathrm{R}_m^j$	$\xrightarrow{k_{\mathrm{td}ij}^\mathrm{l}}$	$\mathrm{P}_n + \mathrm{P}_m$
Precipitation	$\mathrm{R}_\eta^{\mathrm{l}j}$	$\longrightarrow$	$\mathrm{R}_\eta^{\mathrm{g}j}$
Gel phase g			
Decomposition	I	$\xrightarrow{k_\mathrm{d}^\mathrm{g}}$	$2\mathrm{I}^\bullet$
Initiation	$\mathrm{I}^\bullet + \mathrm{M}_j$	$\xrightarrow{k_\mathrm{I}^\mathrm{g}}$	$\mathrm{R}_1^i + (j-1)\mathrm{PDB}$
Propagation	$\mathrm{R}_n^i + \mathrm{M}_j$	$\xrightarrow{k_{\mathrm{p}ij}^\mathrm{g}}$	$\mathrm{R}_{n+1}^j + (j-1)\mathrm{PDB}$
Chain transfer to monomer	$\mathrm{R}_n^i + \mathrm{M}_j$	$\xrightarrow{k_{\mathrm{fm}ij}^\mathrm{g}}$	$\mathrm{P}_n + \mathrm{R}_1^j + (j-1)\mathrm{PDB}$
Termination by disproportionation	$\mathrm{R}_n^i + \mathrm{R}_m^j$	$\xrightarrow{k_{\mathrm{td}ij}^\mathrm{g}}$	$\mathrm{P}_n + \mathrm{P}_m$
Cross-linking	$\mathrm{PDB} + \mathrm{R}_n^i$	$\xrightarrow{f_{\mathrm{PDB}i} k_{\mathrm{p}i2}^\mathrm{g}}$	$\mathrm{R}_{n+1}^2 + \mathrm{X}$

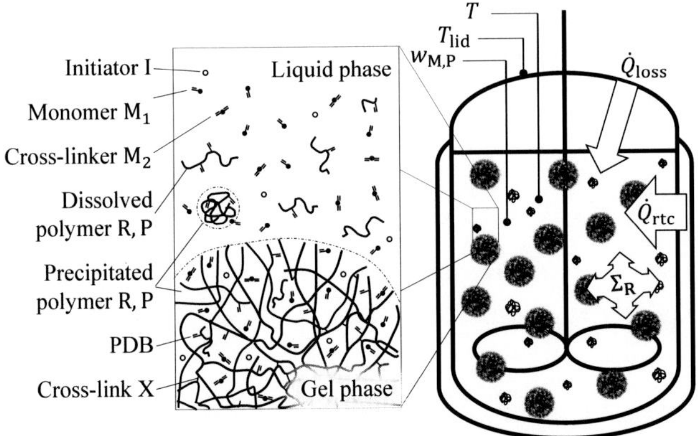

Figure 2.2: Schematic illustration of the two phase system in precipitation polymerization and the present species (left). The precipitation polymerization is performed in an isothermally operated batch reactor. The reactor with the liquid continuous phase and the disperse gel phase is illustrated on the right with the term of the energy balance and the available temperature and mass fraction measurements.

ferred by chain transfer to monomer from an oligomer to a single monomer or cross-linker. This results in the formation of a polymer P_n with chain length n and a radical consisting of a single repeating unit R_1^j and the terminal end j according to the type of the former monomer M_j. Chain transfer to polymer is neglected since it is observed that PVCL does not form polymer networks unless cross-linker is added (Cortez-Lemus & Licea-Claverie (2016)). When a cross-linker is incorporated in a polymer chain by one of the previous mechanisms, the second double bond of the cross-linker is attached to the polymer chain as PDB. The polymerization is terminated when two radicals collide and form inactive polymer.

When the oligomers reach a critical chain length η, they collapse, precipitate, and form the gel phase (g) (Pich & Richtering (2011)). In the gel phase, the reactions described above continue. In addition, the cross-linking reaction occurs when a PDB reacts with another oligomer to form a cross-link X. The two-phase system and the corresponding species listed in Table 2.1 are illustrated in Figure 2.2 on the left.

The mass balances for initiator, monomer and cross-linker as well as balances for the primary chains of oligomer and polymer are derived applying the method of moments. For the formulation of the kinetic model, the following assumptions are made:

- Both liquid and gel phase are considered as reaction loci and both phases are considered to be ideally mixed.
- Separability of reaction kinetics and particle formation applies (Arosio et al. (2011a)).
- As the volume ratio of the two phases changes with proceeding reaction, balances for the reaction species o are formulated in terms of amount of substance instead of concentrations.
- Radical transfer among phases is limited to precipitation, only. Collapse of a dissolved oligomer occurs immediately after obtaining the discrete critical chain length η. The solubility of oligomers in the liquid phase above the critical chain length is neglected. Oligomer below the critical chain length, that are formed in the gel phase (by initiation or chain transfer to monomer), are assigned to the gel phase and do not diffuse into the liquid phase.
- For small compounds (I, M_1, M_2), phase equilibrium between liquid and gel phase is assumed based on partition coefficients P (Arosio et al. (2011a)).
- The occurring reaction is influenced by the type of radical end only and the terminal model can be applied (Alfrey & Goldfinger (1944); Mayo & Lewis (1944)).
- In experiments with pure VCL, it was observed that no stable microgel particles are formed. It can be concluded, that PVCL polymerizes in linear chains and branching has little impact. Hence, the chain transfer to polymer mechanism is neglected.
- The gel phase summarizes all microgel particles, but no distinction among single particles is made. Consequently, the concentrations of the gel phase hold for all particles. A particle can contain several primary chains, and hence several radicals. The growth of the polymer network by cross-linking of primary chains is not considered.
- For the distinction among the two termination mechanisms disproportionation and recombination, measurements of the chain length distribution of the primary radical chains are required. This cannot be determined, since a microgel consists of a single cross-linked macromolecule with multiple primary chains. Thus, termination by disproportionation is assumed as the basis for the formulation of the termination reaction mechanism.
- Potential diffusion limitation (Kiparissides (1996)) for termination reaction is described by the corresponding estimated reaction rate constants. It is assumed that a diffusion limitation exists from the beginning of the gel phase formation (k_{td}^{g} = const.), due to a constantly high polymer concentration in gel phase. Hence, the estimated reaction rate constants are the apparent rate constants.
- Since the molecular structure of BIS is symmetrical, the propagation reaction with the

PDB (cross-linking) has the same reaction rate constant as propagation with the cross-linker. But in contrast to the propagation of an unbound cross-linker, the propagation of a PDB can be sterically limited. This is expressed by an efficiency factor $f_{\mathrm{PDB}i}$ for the propagation with the corresponding terminal end of type i.

- Propagation reactions of PDBs are considered to take place in the gel phase preferably due to higher radical and polymer concentrations. In the liquid phase, cross-linking is neglected, based on the assumption of a short critical chain length resulting in a short residence time of the growing radicals as well as a significantly lower radical concentration in the liquid phase.
- The method of moments is utilized to balance primary chains of oligomer and inactive polymer and to calculate the average number of repeating units l_{avg} of primary chains. However, the calculated moments cannot be applied to calculate a representative number-average molecular weight M_{n} or weight-average molecular weight M_{w} of the microgel particles, as these consist of multiple cross-linked primary chains. Consequently, the second order moments are not considered.

The comprehensive description of the mathematical model is provided in Appendix B.

2.2.2 Energy balance of the batch reactor

For the determination of reaction rate constants from calorimetry, the general formulation of the overall energy balance of the reactor is required. Figure 2.2 illustrates on the right the schematic of an ideally-mixed batch reactor with the largest contributions to the energy balance: the heat transfer rate from reactor jacket to reactor content, $\dot{Q}_{\mathrm{rtc}}$, the heat loss transfer rate through the lid, $\dot{Q}_{\mathrm{loss}}$, and the enthalpy transfer rate due to the exothermic reaction Σ_{R}. The power induced by stirring is assumed negligible. With these terms, the general formulation of the energy balance results in

$$\left(C_{\mathrm{p,ins}} + \sum_{o} m_o c_{\mathrm{p},o}\right) \frac{\mathrm{d}T}{\mathrm{d}t} = \dot{Q}_{\mathrm{rtc}} + \dot{Q}_{\mathrm{loss}} + \Sigma_{\mathrm{R}}. \tag{2.1}$$

The product of the time derivative of the reactor temperature T and the heat capacities of the reactor inserts (stirrer, temperature sensor, turbidity sensor, and baffle) $C_{\mathrm{p,ins}}$ and o reaction species (and solvent) represents the rate of change of the thermal energy. The heat transfer rate $\dot{Q}_{\mathrm{rtc}}$ is directly measured by real-time calorimetry and the heat loss transfer rate $\dot{Q}_{\mathrm{loss}}$ is calculated from the product of the overall heat loss coefficient and the lid surface α, the lid temperature T_{lid} and reactor temperature T:

$$\dot{Q}_{\text{loss}} = \alpha \left(T_{\text{lid}} - T\right). \tag{2.2}$$

Hence, with measurements of $\dot{Q}_{\text{rtc}}$, T and T_{lid}, the enthalpy transfer rate Σ_{R} resulting from the exothermic reaction can be computed from Eq. (2.1).

The fundamental assumption in order to determine the reaction kinetics from Σ_{R} is, that polymerization reaction and particle formation can be considered as independent mechanisms and that precipitation and other particle formation mechanisms are not represented in the energy balance (Arosio et al. (2011b)). Then, Σ_{R} results from the reaction rates of propagation $\mathcal{R}_{\text{p}ij}$ and cross-linking $\mathcal{R}_{\text{PDB}i}$ with $\mathcal{R}_{ij} = k_{ij}c_ic_j$ and the corresponding reaction enthalpies $\Delta H_{\text{R}ij}$. Thereby, there is a relation between reaction rate constants k_{ij} and the energy balance. The corresponding reaction rates for both phases need to be considered, resulting in

$$\Sigma_{\text{R}} = \sum_{j=1}^{2}\sum_{i=1}^{2} -\Delta H_{\text{R}ij}\left(V^{\text{l}}\mathcal{R}^{\text{l}}_{\text{p}ij} + V^{\text{g}}\left(\mathcal{R}^{\text{g}}_{\text{p}ij} + (j-1)\mathcal{R}^{\text{g}}_{\text{PDB}i}\right)\right). \tag{2.3}$$

Eq. (2.3) shows that Σ_{R} is the summation of ten propagation reaction rates. From the energy balance, reaction kinetics can be determined only as an average for both phases. To separate the contribution of each phase to the overall polymerization process, additional information is required. This information is determined from online and offline experiments and quantum mechanical calculations.

2.2.3 Experiments

All microgel synthesis experiments were performed in a Mettler Toledo RT1su reaction calorimeter with a triple-walled 0.5 L RTcal glass reactor operated in isothermal control mode. Measurements of T_{lid}, T, and $\dot{Q}_{\text{rtc}}$ are provided in 2 s intervals. Simultaneously, the mass fractions of monomer and polymer are measured by in-line Raman spectroscopy in 33 s intervals (Meyer-Kirschner et al. (2016)).

For the determination of reaction kinetics, five variations of the experiment are performed. For the pure compound reaction kinetics, first, 4.43 g (31.827 mmol) VCL and second, 1.2266 g (7.956 mmol) of pure BIS are polymerized. In both experiments, no stable microgels were formed. For the copolymerization reaction, three variations of the BIS content were made: 0.06 g (0.389 mmol), 0.12 g (0.778 mmol), and 0.24 g (1.557 mmol) BIS to 4.43 g VCL (1.2 mol-%, 2.5 mol-%, and 5.0 mol-% BIS). 0.0443 g (0.122 mmol) Cetyltrimethylammoniumbromide is added for stabilization of the particles. The water surface is purged with nitrogen for 30 min to remove oxygen before 0.1 g (0.369 mmol) 2,2′-Azobis(2-

methylpropionamidine) dihydrochloride is added for thermal initiation. After initiation, the progress of reaction is measured for 1.5 h to 2 h before the experiments are terminated. At the time of termination, measurements are in steady state again. All experiments are performed under same reaction conditions in a $3 \cdot 10^{-4}\,\mathrm{m}^3$ volume of water at 343 K and ambient pressure. All experiments with VCL are repeated trice, averaged and the standard deviation is calculated. The polymerization of pure BIS is performed once only. Finally, after synthesis, the microgels are dialyzed and characterized by Dynamic Light Scattering (DLS) to determine the hydrodynamic radius.

After initiation, a short inhibition phase ($10-25$ s) is observed before the polymerization begins. The inhibition results from remaining oxygen in the liquid phase. A measurement of the oxygen concentration (303 K) showed that some oxygen remains dissolved in water, when only the atmosphere above the liquid phase is purged. The delay of the polymerization start through inhibition is evaluated for each experiment individually and the inhibition phase is removed for the comparison of experimental data and model.

Further, it is observed that the lid temperature rises with increasing T. Hence, there is a considerable heat loss $\dot{Q}_{\mathrm{loss}}$ through the lid of the reactor which is required for the precise calculation of Σ_{R} from calorimetry measurements by Eq. (2.1). Hence, the product of heat transfer coefficient and lid area, α, is determined from an additional experiment under comparable synthesis conditions. $3 \cdot 10^{-4}\,\mathrm{m}^3$ water is heated to $T = 343\,\mathrm{K}$ while the T_{lid}, $\dot{Q}_{\mathrm{rtc}}$, and T are measured simultaneously until steady state is obtained. Then, α is calculated with Eqs. (2.1)-(2.2) and measurements from steady state (cf. Appendix B.4). With α determined individually, $\dot{Q}_{\mathrm{loss}}$ can be determined and thus, Σ_{R} can be calculated for the performed experiments. Further, the energy balance can be easily conveyed to non-isothermal process operation.

In addition to the microgel syntheses, the critical chain length is investigated experimentally using the lower critical solution temperature (LCST) of linear PVCL. For the reaction temperature of 343 K, the critical chain length at which the collapse of the polymer chain occurs is determined to be $\eta = 12$ repeating units.

2.2.4 Quantum mechanical calculations

The model for precipitation polymerization as described above includes two initiator efficiencies (f_{I}), two cross-linking efficiencies (f_{PDB}), three partition coefficients (P), four reaction enthalpies (ΔH_{R}), and 22 reaction rate constants (k). Most of these parameters affect Σ_{R} directly or indirectly. Therefore, the parameter values cannot or only inaccurately be determined from the measurements. To reduce the number of parameter values to be determined from experimental data using Eq. (2.3), quantum mechanical (QM) calculations, transition state theory and COSMO-RS (Klamt (1995)) are employed. As described in Kröger et al. (2017), the rate constants for chain propagation reactions $k^{\mathrm{l}}_{\mathrm{p}ij}$ and chain transfer to monomer reactions $k^{\mathrm{l}}_{\mathrm{fm}ij}$ are determined for the liquid phase as well as the corresponding reaction enthalpies $\Delta H_{\mathrm{R}ij}$. For the gel phase, the kinetic constants are calculated in dependency of surrounding polymer concentrations under the assumption of 90 w-% polymer to 10 w-% water (cf. Wu et al. (1994) for PNIPAM-based microgels). Further, the partition coefficients for monomer and cross-linker are computed using COSMO-RS under the assumption of phase equilibrium (Klamt (1995); Loschen & Klamt (2014)). It has been shown that for charged molecules, parameters calculated from solvation models (i.e. COSMO-RS) are more likely to have large errors (Cramer & Truhlar (2008)). Hence, the partition coefficient and the reaction rate constants for the initiator are not predicted. Instead, the decomposition rate constant provided by the manufacturer [1] is used. Initiator efficiency factors as well as its partition coefficient are estimated. The parameter values used for simulation are listed in the Appendix Table A.1 and Table A.4. Employing this approach, the number of unknown parameter values is reduced from 31 to 11: four efficiency parameters, one partition coefficient and six apparent kinetic constants for termination reactions.

2.2.5 Parameter estimation

The cross-linking copolymerization model with parameters from QM calculations is compared to calorimetry and Raman measurements for evaluation. It is implemented in gProms Model Builder Version 4.2.0 [2]. The unknown kinetic parameters are estimated by the standard maximum likelihood method of gProms Model Builder. The unknown parameter values are estimated from measurements of reactor temperature T, mass fractions of monomer w_{M_1} and polymer w_{P} while T_{lid} and $\dot{Q}_{\mathrm{rtc}}$ measurements serve as input variables to the model. The experimental data of reaction calorimetry and Raman spectroscopy consistently show the most pronounced dynamic behavior within the first 500 s after initiation and the subsequent inhibition phase. Hence, this interval is used for parameter estimation. Moreover, experimental and simulation data are compared for this interval.

The parameters are estimated in a sequential approach to reduce the number of simultaneously estimated parameter values: First, homopolymerization parameter values are estimated based on the polymerizations of pure VCL and pure BIS and second, cross-reaction kinetic parameters are estimated from experiments with varying concentrations of BIS.

[1] Wako Pure Chemical, Ltd. `https://www.wako-chemicals.de/en/product/v-50`, accessed 30.03.2017

[2] Process Systems Enterprise, gProms, www.psenterprise.com/gproms 1997-2020

2.3 Results and discussion

In this section, the simulation results of the fitted model are compared to experimental measurements of the respective experiments. The simulation will be evaluated first for the homopolymerization of VCL and BIS, respectively. Then, results for cross-linking copolymerization of VCL with BIS are discussed. The determined reaction kinetics are then combined with the results of microgel characterization to predict the internal structure of the synthesized microgels.

2.3.1 Homopolymerization

First, the parameters $P_{\rm I}$, $f_{\rm I}$, $k_{\rm td11}$ and $k_{\rm td22}$ for both phases are estimated based on measurements of pure VCL and pure BIS polymerization. The cross-linking reaction is accounted for by estimation of the ratio parameter $f_{\rm PDB2}$ for the reaction of pure BIS, only.
Figure 2.3 shows the comparison of $\Sigma_{\rm R}$ derived from experiments and simulation. The experimental $\Sigma_{\rm R}$ is calculated from Eq. (2.1) employing measurements of the temperature gradient and $\dot{Q}_{\rm rtc}$. The simulated $\Sigma_{\rm R}$ is calculated from the reaction kinetics by Eq. (2.3). By employing

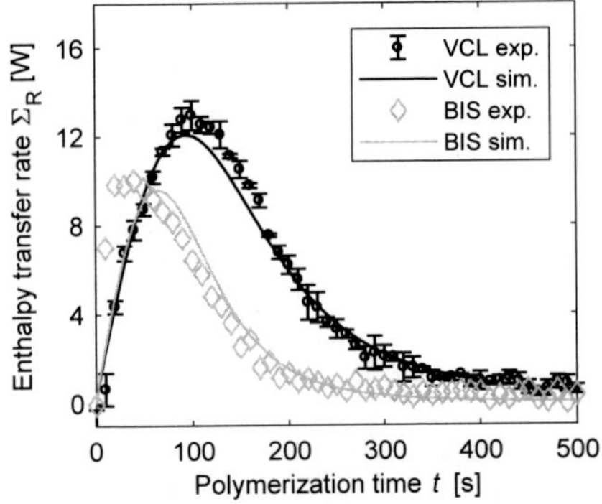

Figure 2.3: Comparison of simulated and experimental $\Sigma_{\rm R}$ for VCL (black) and BIS (grey) polymerization. For improved readability, experiments are shown in 10 s interval only. To calculate $\Sigma_{\rm R}$ from measurements of T, $\dot{Q}_{\rm rtc}$ and $T_{\rm lid}$, the same parameter values are used as in the model. The bars denote the standard deviation of three repetitions of the experiment with VCL.

the kinetic constants and reaction enthalpy from QM calculations, good agreement of simulation and experiments can be obtained. The reaction time for BIS is slightly faster than for VCL. This corresponds to the observation that the kinetic constant for BIS, ($k_{p22} = 8.1\,\mathrm{m}^3\,(\mathrm{mol\,s})^{-1}$), is higher than for VCL, ($k_{p11} = 1.34\,\mathrm{m}^3(\mathrm{mol\,s})^{-1}$). The higher kinetic constant even compensates the low initial BIS concentration ($c_{M_2} = 26.52\,\mathrm{mol\,m}^{-3}$), which is only 25 % of the initial VCL concentration ($c_{M_1} = 106.09\,\mathrm{mol\,m}^{-3}$). The low BIS concentration is necessary to prevent high viscosity, consecutive diffusion limitation, or polymer layers on the surfaces of the reactor. Otherwise, these effects can affect the estimates for the apparent reaction kinetics.

Figure 2.4 depicts the Raman measurements for the experiments described above in compari-

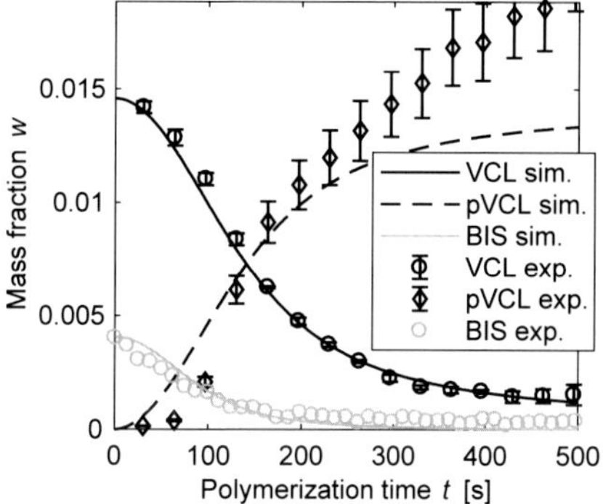

Figure 2.4: Comparison of simulation with Raman measurements of the polymerization of VCL and PVCL (black) and polymerization of BIS (gray). The bars denote the standard deviation of three repetitions of the experiment. The deviation for high polymer mass fractions results from a structural error in measurement. The simulation respects the law of mass conservation and gives a better prediction of the actual polymer content.

son to the simulated mass fractions. The bars depict the standard deviation for three repetitions of the experiment and hence account for the statistical uncertainty only. For both, monomer VCL and cross-linker BIS, there is good agreement of simulation and measurements. The most distinct deviation is observed for the polymer mass fraction for later polymerization time. The model respects the law of mass conservation and the calculated mass fraction converges to a constant value, which should also be observed experimentally. However, the experiments show a structural measurement error in terms of an overestimation of the polymer mass fraction (Meyer-Kirschner et al. (2016)). The reason for this error is the weak Raman signal of the polymer. In addition, the Raman signal is not characteristic but it overlaps with the more significant monomer Raman signal, which makes its determination error-prone. Hence, for long polymerization time, the simulation gives a better approximation of the true polymer mass fraction than the experiments do.

2.3.2 Copolymerization

With the kinetic constants from QM calculations and the previously determined pure component reaction constants, the remaining parameters k_{td12}, f_{PDB1}, f_I and P_I are estimated from experimental data for three different cross-linker concentrations. Comparing the calorimetry measurements by means of Σ_R in Figure 2.5, an influence of the cross-linker concentration can be observed. With increasing cross-linker concentration, a second peak appears. While the maximum of the first peak increases with increasing cross-linker concentration, the maximum of the second peak decreases slightly.

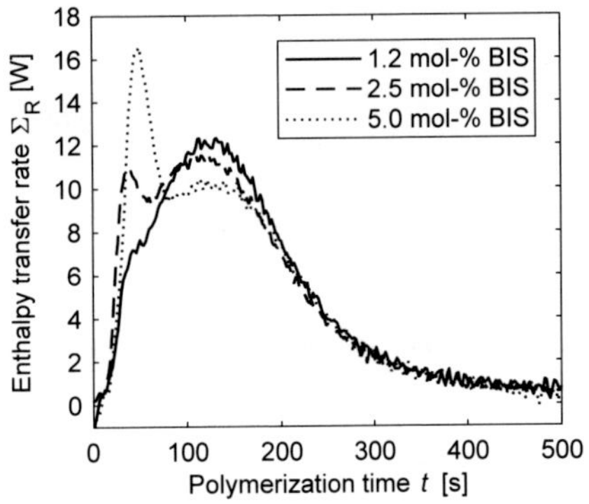

Figure 2.5: Σ_R calculated from experiments according to Eq. (2.1) for synthesis with varying BIS concentrations. Parameters for energy balance equal those of simulation.

Figure 2.6 depicts the simulated Σ_R corresponding to the experiments in Figure 2.5. Σ_R shows the identical behavior of the two peaks regarding the cross-linker concentration. As this effect can be reproduced with the kinetic constants from QM calculations, it can be linked to the cross-reactions of VCL and BIS. The reaction rate constants of the cross-reactions are 10 to 100 times larger than the pure component reaction constants (Kröger et al. (2017)). This compensates the low ratios of cross-linker to monomer concentrations. Hence, the reaction rates $\mathcal{R}_{p12}$ and $\mathcal{R}_{p21}$ are comparable in scale to the reaction rate $\mathcal{R}_{p11}$ and their effect on Σ_R

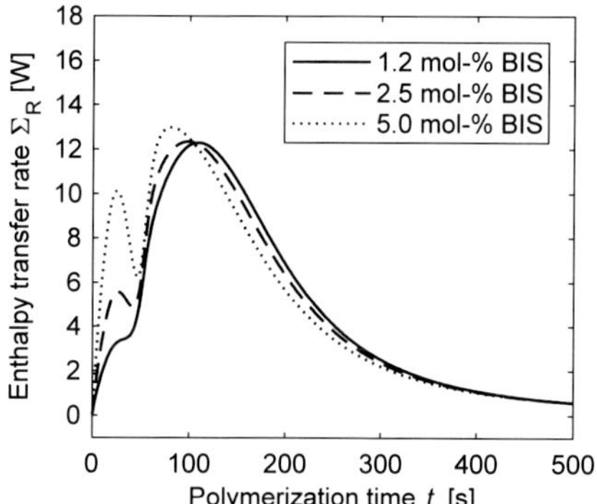

Figure 2.6: Simulated product of reaction rates and reaction enthalpies Σ_R for synthesis with varying cross-linker concentrations.

can be observed. When the cross-linker is consumed, the first peak decreases and the controlling reaction is propagation of pure VCL, resulting in the second peak. When comparing simulation (Figure 2.6) to the experiments (Figure 2.5), the first peak is located 25 s earlier than in the experiment. This indicates that the cross-linker is consumed faster in simulation.

In Figure 2.7, the calculated mass fractions for VCL and PVCL are compared to Raman spectroscopy measurements. For the monomer mass fraction for both experiments, simulation and measurements are in very good agreement. The experiment with 5.0 mol-% BIS and its simulation reveal a slightly faster decrease of w_{M_1} right after initiation, though the further progress of w_{M_1} for both experiments, 1.2 mol-% BIS and 5.0 mol-% BIS, show no deviation. In general, the simulation of w_P reproduces the trend of measurements well. However, at the beginning of the polymerization, simulated w_P shows a faster increase, and at the end, a lower final polymer mass fraction. Again, the overestimated experimental w_P towards the final polymerization time results from the structural measurement error (Meyer-Kirschner et al. (2016)), while the simulation respects the law of conservation of mass and gives a better approximation of w_P. The same applies for the beginning of the polymerization. The measurement points before 70 s, w_{M_1} decreases observably, while no w_P is measured. This can be explained by the weak Raman signal of PVCL, which leads to the underestimation of the present polymer mass fraction at low polymer concentrations. However, the simulated polymer mass fraction includes the sum of reacted monomer as well as reacted cross-linker, and it increases accordingly to the decrease in the monomer mass fraction. Hence, the simulation predicts the actual present polymer mass fraction.

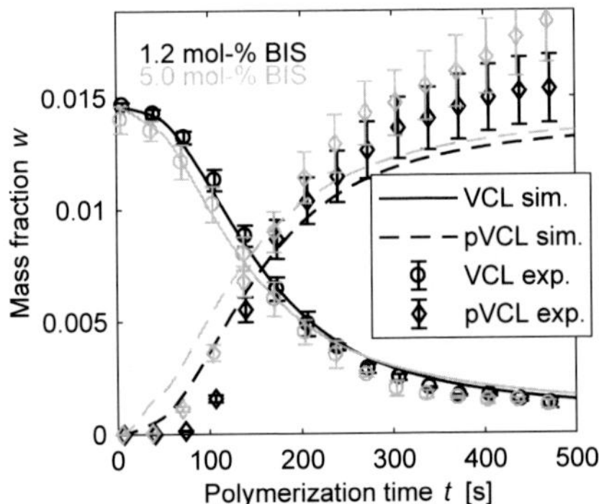

Figure 2.7: Comparison of simulation with Raman spectroscopy measurements (Meyer-Kirschner et al. (2016)) for experiments with 1.2 mol-% and 5.0 mol-% BIS. The bars denote the standard deviation of three repetitions of the experiment. The deviation for high polymer mass fractions results from a structural error in measurement. The simulation respects the law of mass conservation and gives a better prediction of the actual polymer content.

The low cross-linker concentration makes determination of the cross-reaction rate constants difficult. Reaction calorimetry measurements reveal a distinct dependency on the cross-linker concentration, but the calculated Σ_R provides only the overall polymerization progress (cf. Eq. (2.3)) and hence the total consumption of monomer and cross-linker. Although the cross-linker concentration is too small for a direct measurement by the employed Raman spectroscopy, the monomer concentration can be measured with high accuracy (Meyer-Kirschner et al. (2016)). The combination of both allows for the calculation of the cross-linker consumption and hence the validation of QM reaction rate constants as well as the estimation of missing cross-reaction rate constants.

For both, homopolymerization and copolymerization reactions, the estimates for termination reaction rate constants in the gel phase are significantly smaller than in the liquid phase (cf. Table A.5). For the liquid phase, parameter values are in the order of magnitude that has been reported for solution polymerization (Odian (2004)). Despite the low critical chain length of 12 repeating units, the termination reaction in the liquid phase has an impact on the simulation results, since a faster consumption of monomer is predicted when termination in the liquid phase is neglected. A possible interpretation for the low parameter estimates in the gel phase would be the diffusion limitation of termination. However, the estimates for the termination reaction rate constants are highly correlated with the initiator efficiencies and the initiator partition coefficient and hence need to be treated with caution. An identifiability analysis remains to be performed to provide some indication of the reliability of the parameter estimates.

Nonetheless, QM calculations already provide great value to polymerization modeling of the multi-component system. Jung et al. (2019a) performed a parameter estimation with corresponding identifiability analysis for a significantly reduced copolymerization model derived from the model presented in this chapter. Therein, it is assumed that no parameter values from QM calculation are given. The identifiability analysis showed that specifically parameter values for cross-linker propagation reactions cannot be determined from experimental data with certainty. Hence, incorporation of parameter values from QM calculations is an important tool for process modeling of complex polymerization systems. The successful incorporation of parameter values from QM calculations in process models is also demonstrated in a case study for cross-linked PVCL-PNIPAM-based microgels provided in Appendix C.

2.4 Prediction of microgel properties

The model is used to give insight on the contribution of each phase to the overall polymerization reaction and into the cross-linking reaction. The combination of both allows conclusions on the internal particle structure regarding the cross-link distribution.
Figure 2.8 shows the amount of unreacted cross-linker, PDBs and formed cross-links for the simulation with 2.5 mol-% BIS. As the cross-linker is consumed, the number of PDBs and cross-links increases. The maximum number of PDBs is obtained when the cross-linker is fully consumed. Then, no further PDBs are formed but consumed only to form cross-links. The simulation shows that, though the cross-linker is consumed comparably fast, the cross-linking reactions occur throughout the entire polymerization time. At the end of the simulation, only 60 % of the PDBs formed a cross-link, though the cross-linking reaction is not completed.

The experimental determination of the gel volume under reaction conditions is difficult and due to the remaining water in the polymer network most likely error-prone. On the other hand, the model provides an approximation of the gel formation (cf. Appendix B.2). Two mechanisms contribute to mass transfer from liquid to gel phase: precipitation of radical chains of critical length and transfer of monomer driven by phase equilibrium. Both are fully determined by the known reaction rate constants for propagation, the partition coefficients, and the critical chain length. Hence, contribution of the gel phase to the overall monomer consumption can be calculated. Figure 2.9 shows the instantaneous fraction of monomer and cross-linker consumed in the gel phase of the overall consumed monomer and cross-linker, $F_{\mathrm{g,inst}}$ (cf. Eq. (2.4)), which is calculated from the reaction rates ($\mathcal{R}^{\mathrm{l,g}} \mathrel{\hat=} \sum_i^2 \sum_j^2 \mathcal{R}_{ij}^{\mathrm{l,g}}$ of the corresponding

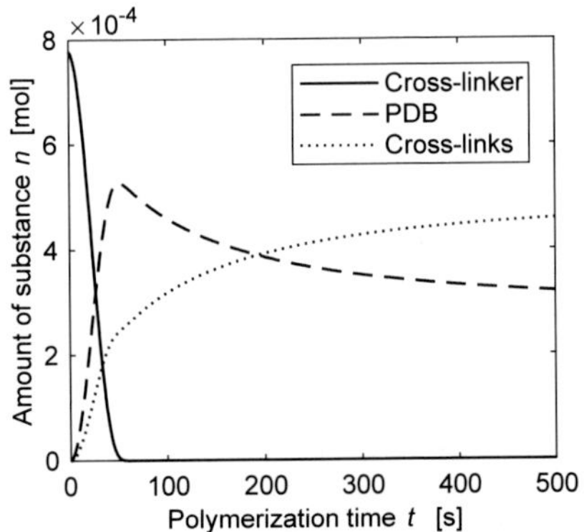

Figure 2.8: Simulated profiles for cross-linker, pendant double bonds and cross-links (2.5 mol-% BIS).

reactions).

$$F_{\mathrm{g,inst}} = \frac{V^{\mathrm{g}}\left(\mathcal{R}_{\mathrm{p}}^{\mathrm{g}} + \mathcal{R}_{\mathrm{fm}}^{\mathrm{g}} + \mathcal{R}_{\mathrm{I}}^{\mathrm{g}}\right)}{V^{\mathrm{l}}\left(\mathcal{R}_{\mathrm{p}}^{\mathrm{l}} + \mathcal{R}_{\mathrm{fm}}^{\mathrm{l}} + \mathcal{R}_{\mathrm{I}}^{\mathrm{l}}\right) + V^{\mathrm{g}}\left(\mathcal{R}_{\mathrm{p}}^{\mathrm{g}} + \mathcal{R}_{\mathrm{fm}}^{\mathrm{g}} + \mathcal{R}_{\mathrm{I}}^{\mathrm{g}}\right)} \cdot 100\% \tag{2.4}$$

$F_{\mathrm{g,inst}}$ is applied as a measure for the monomer and cross-linker consumption in the gel phase at any polymerization time. Immediately after initiation, more than 90 % of the monomer and cross-linker is consumed in the gel phase. This is compared to the fraction of gel volume of the total control volume. Although its final volume makes only 1 % of the total volume, the majority of monomer and cross-linker is consumed in the gel phase. At the end of the simulation, 98 % of the cross-linker and 89 % of the monomer are reacted in the gel phase. By this time, 100 % of the cross-linker and 90 % of the monomer is consumed. Despite the decreasing effect of the polymer concentration on the propagation rate constants (cf. Appendix A.1), a similar effect to what has been observed experimentally for system of *N*-Vinylpyrrolidone (Uhelská et al. (2014)), the monomer and cross-linker consumption is higher in the gel phase. Both of the assumed mass transfer mechanisms contribute to this. The short critical chain length of 12 repeating units leads to fast precipitation and hence results in a high radical concentration in the gel phase. Partition coefficients from QM calculations predict a higher monomer and cross-linker concentration in the gel phase. This causes monomer and cross-linker mass flows from the liquid to the gel, which is a multiple of the mass flow from precipitation. Hence, the simulation shows that the gel phase is the primary reaction locus.

In this context, note that the partition coefficients express a higher hydrophobicity of the

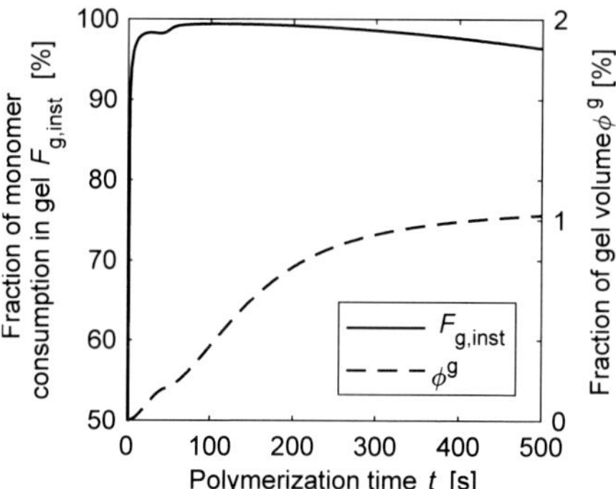

Figure 2.9: Contribution of gel phase as reaction locus in comparison to the fraction of gel phase in the solution. Simulation of experiment with 2.5 mol-% BIS.

monomer than the cross-linker. The ratio of cross-linker to monomer concentration is lower in the gel phase than averaged over both phases. This is the reverse effect to the fast cross-propagation rate constants, and it delays the consumption of the cross-linker. As the experiments show a faster consumption of the cross-linker, this indicates that the reaction rate constants cause the inhomogeneous cross-linker consumption.

The formation of the gel phase is combined with the reaction rates to predict the average internal structure of a particle regarding the local distribution of cross-links. For monitoring the internal cross-linking, the instantaneous cross-linking rate fraction $F_{X,inst}$ is calculated from the ratio of cross-linking reaction rates to the sum of reaction rates of other growth reactions:

$$F_{X,inst} = \frac{\mathcal{R}^{g}_{PDB}}{\mathcal{R}^{g}_{p} + \mathcal{R}^{g}_{fm} + \mathcal{R}^{g}_{I} + \mathcal{R}^{g}_{PDB}} \tag{2.5}$$

In Figure 2.10 on the left, $F_{X,inst}$ is plotted against the instantaneous particle radius, which defines the maximal volume, within which the cross-linking reaction occurs, to predict the cross-link distribution for the experiments with 1.2, 2.5, and 5 mol-% BIS. The instantaneous particle radius is thereby calculated from the simulated gel volume and the final particle radius in collapsed state from DLS measurements. This is based on the assumption that particle nucleation occurs in a short period after initiation and thus of particle number remains constant. Then, the particles grow likewise by absorption of monomer, cross-linker, and precipitated polymer, resulting in the observed monodisperse particles (cf. Appendix B.3).

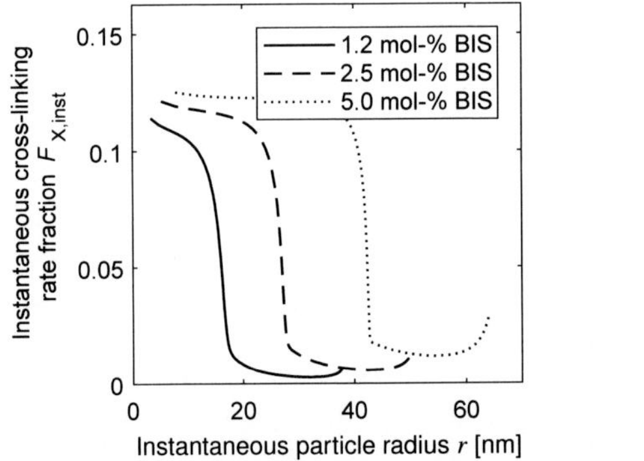

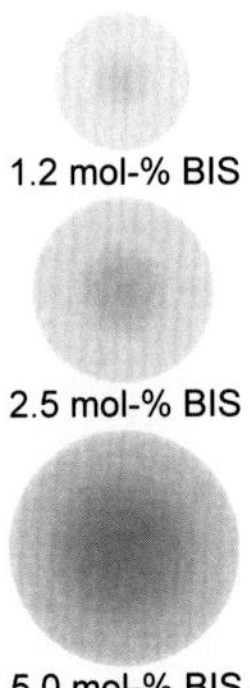

Figure 2.10: Cross-linking density in microgels for different initial cross-linker concentrations. Left: Simulated instantaneous cross-linking mole fraction compared to corresponding calculated instantaneous particle radius for different cross-linker concentrations. Right: Illustration of cross-linked particles: The size of the particles increases with increasing cross-linker concentration (Mean hydrodynamic radius in collapsed state (DLS measurements at 323 K): 1.2 mol-%: 38 nm; 2.5 mol-%: 50 nm; 5.0 mol-%: 64 nm). All particles show a core-shell structure with a densely cross-linked core, marked dark gray, and a sparsely cross-linked shell.

For all simulated cross-linker concentrations, $F_{X,inst}$ shows a high density in the center, which decreases step-like. The radius for the change from high to low cross-linking mole fraction spheres increases with the cross-linker concentration. Also, the minimal $F_{X,inst}$ increases with increasing cross-linker concentration. Towards the maximal particle radius, $F_{X,inst}$ increases again. At this late stage of the polymerization, the propagation reaction rates are low as monomer and cross-linker are mostly consumed and the particle growth comes to hold. But the cross-linking reactions continue as PDBs are still available (cf. Figure 2.8) and hence, the cross-linking rate fraction increases.

According to the assumption of an ideally-mixed gel phase, the cross-linking reactions occur homogeneously throughout the gel particle. Hence, cross-linking in the center of the particle is described by the accumulation of $F_{X,inst}$ over the entire reaction time. The resulting particle cross section is schematically illustrated in Figure 2.10 on the left. The intensity of the gray corresponds to the cross-link distribution of the particle spheres: highly cross-linked spheres are represented by dark gray and light gray represents spheres with sparse cross-linking. All particle cross sections show a highly cross-linked core and a lighter cross-linked shell. The radius of the cross-linked core and the cross-link density of the shell both increase with increasing initial cross-linker concentration. The inhomogeneous internal structure of VCL/BIS-based microgels which is predicted by the model has also been observed experimentally (Schneider et al. (2014)).

2.5 Conclusion

In this chapter, the synthesis of PVCL-based microgels by precipitation polymerization was modeled as a two-phase cross-linking copolymerization. The formulation of the energy balance allowed linking the reaction kinetics to calorimetry measurements. Measurements of chain length-dependent solubility of the polymer were employed to describe the formation of a two-phase system by precipitation. Quantum mechanical calculations were used to calculate partition coefficients, reaction rate constants, and the corresponding reaction enthalpies. This reduced the number of unknown parameter values considerably. The model is validated with measurements from calorimetry and Raman spectroscopy and showed, disregarding the observed oxygen inhibition, qualitatively good agreement. While the measurements for validation were averaged over both liquid and gel phase, the model enabled the prediction the gel phase volume, the contribution of each phase to the overall reaction, and hence allowed the prediction of the reaction locus. Further, based on the determined reaction kinetics, the cross-linker incorporation as well as the cross-linking formation was calculated. The prediction of the gel phase was combined with the calculation of cross-linking to derive the average internal structure of a microgel particle regarding its cross-linking distribution.

Three central observations can be concluded: First, the comparison of simulation and measurements shows that the cross-linking copolymerization is a well-suited model for the synthesis of PVCL-based microgels with the cross-linker BIS. Second, calorimetry and Raman spectroscopy are well-suited measurement techniques to provide valuable insight into the reaction kinetics of copolymerization even for low cross-linker concentrations. Last, the parameters from QM calculations can be employed successfully for the simulation of both pure VCL and pure BIS polymerization process as well as VCL and BIS copolymerization processes.

The use of QM calculation enables the reduction of the number of unknown parameter values from 31 to 11. Nevertheless, some of the estimated parameters are highly correlated, which limits the determination of the exact parameter values of the reaction kinetics from the available measurements. Hence, a comprehensive identifiability analysis remains to evaluate the reliability of the estimated parameter values and determine the prediction capabilities of the model.

Besides the reaction kinetics, control of particle size and polydispersity of the particles is of great interest for the synthesis of microgels with specific properties. The contribution of the different particle formation mechanisms, e.g., absorption of precursor particles or aggregation of particles, is not yet fully understood and will be addressed in Chapter 3. This will enable the description of the key features of microgels as a first step towards process design for the synthesis for functional PVCL-based microgels.

Chapter 3

Particle size distribution model for microgel synthesis

In the model developed in Chapter 2, the polymerization reactions in terms of conversion of the monomers were investigated. Though the model can provide a rough prediction of the microgel structure based on the gel phase growth, it is not capable of a prediction of the microgel size itself. In the following, modeling the growth of microgels will be addressed through the example of PVCL-based microgels. Therein, the focus is set on the two properties to describe growth being microgel size and size distribution.

3.1 Introduction and literature review

Microgels synthesized by batch precipitation polymerization show typically a narrow particle size distribution (Pich & Richtering (2011)). Following the described formation mechanism in Chapter 2, precursor particles are formed when dissolved oligomers collapse due to a lower solubility of oligomers with higher chain lengths. At this stage, the homogeneous system with dissolved oligomers changes into a heterogeneous system with a continuous aqueous phase and disperse polymer particles. The precursor particles continue to grow until the final microgels are obtained. Growth of the polymer particles thereby is a combination of several simultaneous mechanisms illustrated in Figure 3.1: Polymerization by absorption of monomer and cross-linker and chain propagation reactions in the microgels, entry and desorption of oligomers as well as coagulation among the growing microgels.

The different growth mechanisms determine the characteristics of the particle size distribution. However, the contributions of the individual growth mechanisms are not fully understood so far. A better understanding of growth mechanisms will enable the targeted synthesis of pre-

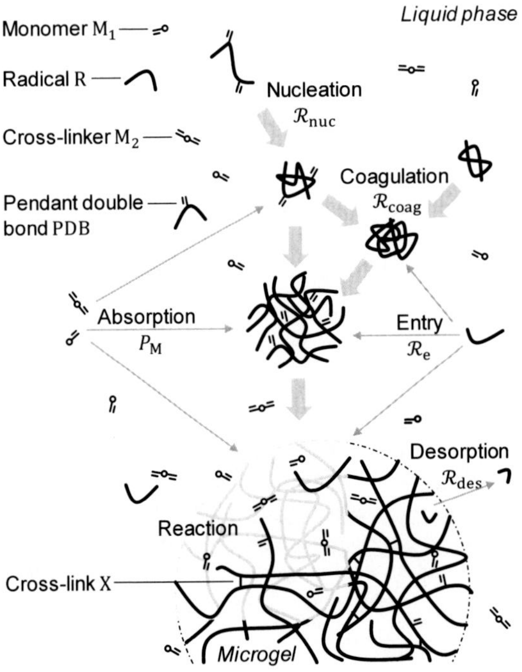

Figure 3.1: Illustration of different particle growth mechanisms contributing to the microgel growth in precipitation polymerization: Nucleation (rate $\mathcal{R}_{nuc}$), monomer absorption (partition coefficient P_M), radical ab- and desorption (rates $\mathcal{R}_e$ and $\mathcal{R}_{des}$) and particle coagulation (rate $\mathcal{R}_{coag}$)

defined microgel sizes by tailored process operation. For this purpose, process modeling and simulation is a valuable tool to complement experimental investigations.

Several experimental studies address the effects of different operation conditions of the PVCL-based microgel synthesis on the obtained microgel sizes. Imaz & Forcada (2008b) provide a fundamental study of the impact of concentration of monomer, cross-linker, initiator, surfactant as well as temperature in emulsion polymerization. They report that increasing cross-linker and initiator concentrations lead to increasing microgel diameters, while an increase of the surfactant concentration results in smaller diameters. In a consecutive study, the effect of different cross-linker types, BIS and the bio-compatible cross-linker poly(ethylene glycol) diacrylate (PEGDA), and their respective concentrations on the particle size are compared (Imaz & Forcada (2008a)). Schneider et al. (2014) characterize the particle size in relation to the cross-linker concentration for precipitation polymerization, with particular focus on the radial heterogeneous cross-link distribution within the particles and the resulting swelling behavior. These studies emphasize the multitude of influencing factors of the final microgel sizes.

So far, only few models have been proposed for the microgel synthesis. The focus of these contributions is rather the internal copolymer composition of microgels such as cross-linking density, while little attention is paid to particle size (distribution). Hoare & McLean (2006b) propose a kinetic model for the the prediction of the internal microgel structure of poly(*N*-isopropylacrylamide) (PNIPAM)-based microgels. The study correlates the solution polymerization reaction rates of NIPAM with several comonomers with the local polymer composition and further employs this approach to adjust the microgel internal copolymer composition (Hoare & McLean (2006a)). With the assumptions of a monodisperse particle size distribution and the lacking onset of gelation, solution polymerization kinetics can be applied to describe a surfactant free emulsion polymerization of PNIPAM-based microgels. Similarly, Acciaro et al. (2011) employ the solution polymerization rate constants of PNIPAM-based microgels for information of the internal distribution as microgel growth is assumed to occur by polymerization on the surface of the particles. Based on reaction rates of the solution copolymerization, a feeding strategy is determined to synthesize homogeneously cross-linked microgels. Likewise, assuming solution polymerization mechanism, Virtanen et al. (2015) provide empirical equations to predict the particle number density and hence, particle size of PNIPAM-based microgels in dependence of the monomer and initiator concentration.

In contrast to these solution polymerization based models, the model described in Chapter 2 describes the precipitation polymerization of PVCL-based microgels as a two-phase system. The model distinguishes among an aqueous liquid and a polymer-rich gel phase. Mass transfer among the phases is described by precipitation at critical chain length. Comparing the impact of the polymerization reactions in gel to liquid phase, the gel phase is identified as the primary reaction locus. The growth of the gel phase related to the cross-linking reaction provides a prediction of the internal cross-linking distribution.

Concurrently with the development of the model presented in this chapter, Jung et al. (2019b) presented a predictive model to describe the microgel growth during precipitation polymerization. The simplified process model for polymerization reactions (Jung et al. (2019a)), which is build upon the model presented in Chapter 2, is extended to capture the number of microgel particles. The particle number results from initiation and coagulation. With the particle number and the total gel phase volume, the average particle radius is calculated. The coagulation rate constant is therein determined from experimental data, however, does not take into account different aggregation probabilities for microgels of different size.

All of the previous models for microgel synthesis share in their prediction of particle growth the assumption of a uniform particle size. The narrow particle size distribution is mechanistically explained by a short particle formation phase, after which the formed microgels are considered to grow evenly. But this assumption lacks a mechanistic description of the microgel formation, and in particular prevents the description of the period of precipitation and coagulation, the number of formed microgels and their size and particle size distribution. This information can be captured by population balance equation (PBE) models.

PBE models are widely applied for emulsion polymerization (Kiparissides (2006); Sheibat-Othman et al. (2017); Thickett & Gilbert (2007); Vale & McKenna (2005)). PBE can be distinguished into 0-1 models and pseudo-bulk-models, based on the number of radicals present in a polymer particle. In 0-1 models, only polymer particles with either no or a single radical exist as the radical undergoes immediate termination in presence of a second radical (Coen et al. (1998); Crowley et al. (2000); Sajjadi (2003)). Hence, they are generally recommended for disperse systems with small particle sizes, unless the radical propagation is fast (Sheibat-Othman et al. (2017); Thickett & Gilbert (2007)).

In contrast, pseudo-bulk models have no limitation regarding the number of radicals. Instead, an average number of radicals for particles of the same radius is assumed. The assumptions of pseudo-bulk models are often controversially discussed due to lack of accuracy for small particles (Thickett & Gilbert (2007)), which has a large impact on the prediction of particle formation, and alternative modifications have been proposed to overcome this issue (Coen et al. (2004)). However, the pseudo-bulk model is the most general formulation and hence, more suited to describe the entire conversion range (Sheibat-Othman et al. (2017)). Araújo et al. (2001) presented a pseudo-bulk model for copolymerization with coagulation. Therein, colloidal stability of the polymer particles is expressed in terms of stable particles, which are unable to undergo coagulation with other stable particles. Immanuel et al. (2002) provide a comprehensive first principle pseudo-bulk model for the seeded emulsion polymerization of vinyl acetate with butyl acrylate. In addition to micellar nucleation, homogeneous nucleation due to precipitation is considered while coagulation is neglected. In a subsequent contribution, they extend the pseudo-bulk model by coagulation under the impact of sterical stabilizer (Immanuel et al. (2003)). Kiparissides et al. (2002) apply a pseudo-bulk model to describe the impact of oxygen on the particle size distribution of vinyl chloride-based latex particles. Brunier et al. (2017) apply the pseudo-bulk model for emulsion polymerization. The focus of their contribution is on ongoing diffusion limitation within the gel particles, beginning with the gel effect and leading to full stagnation of the polymerization by the emerging glass effect. The drawback of the lacking accuracy is avoided by neglecting the first interval of emulsion polymerization and coagulation, focusing on preexisting and stable particles. In summary, pseudo-bulk models are probably the most applied models for emulsion polymerization systems.

In this chapter, a pseudo-bulk polymerization model is employed to the synthesis of PVCL-based microgels. Several considerations argue in favor of the pseudo-bulk model for this purpose: First, Wu et al. (1994) and Varga et al. (2001) concluded for PNIPAM-based microgels that more than one radical can be present in a growing microgel. Further, when the termination is not the rate determining step in the overall polymerization progress, pseudo-bulk models are favorable over zero-one models (Thickett & Gilbert (2007)). The termination rate constants in the gel phase determied with the previous model (cf. Chapter 2.2.5) are low, as is usually the case with diffusion-controlled termination, and hence, polymerization is rate determining. Finally, more explicit models such as 0-1-2, 0-1-2-3 or hybrid 0-1-pseudo-bulk models require population balances for more particle species, distinguished by their respective number of radicals (Coen et al. (2004); Sajjadi (2009); Vale & McKenna (2009)). This comes with high computational cost, especially when parameter estimations are performed. Hence, the pseudo-bulk model comes in handy for this first-time investigation.

From the well-established emulsion polymerization models, mechanisms such as the homogeneous nucleation by precipitation of oligomers can be adapted to describe precipitation polymerization. In the following, the first pseudo-bulk model to describe the precipitation copolymerization with internal cross-linking for the synthesis of PVCL-based microgels is presented. The model uses well-established descriptions for the growth mechanisms radical entry, radical desorption and coagulation. A parameter estimation is performed to determine the contributions of radical entry, desorption and coagulation from experimental data. Therein, the central question is whether the model can equally be fitted to the experimental data from reaction calorimetry, while simultaneously achieving the experimental particle size with a narrow particle size distribution. With the estimated parameter values, simulations for different initial initiator and cross-linker concentration as well as reaction temperatures are performed and compared to experimental data to evaluate the qualification of the pseudo-bulk model for this purpose.

3.2 Modeling particle size distribution for precipitation polymerization

The population balance equation (PBE) is the central equation to the pseudo-bulk model:

$$\left.\frac{\partial \mathcal{F}}{\partial t}\right|_{r,t} + \left.\frac{\partial \mathcal{F} v}{\partial r}\right|_{r,t} = \delta_{r_{\text{nuc}},r}\mathcal{R}_{\text{nuc}}(r,t) + \mathcal{R}_{\text{coag}}(r,t), \tag{3.1}$$

Table 3.1: Reaction scheme of free radical copolymerization occuring in both phases (liquid phase l and gel particles g). The indices i, j correspond to the respective species of monomer or terminal end ($i, j = 1$ for VCL and $i, j = 2$ for BIS).

Reaction			
Initiator decomposition[a]	I	$\xrightarrow{k_d}$	$2I^\bullet$
Initiation	$I^\bullet + M_i$	$\xrightarrow{k_{pi}}$	R_1^i
Chain propagation	$R_n^i + M_j$	$\xrightarrow{k_{pij}}$	$R_{n+1}^j + (j-1)PDB$
Chain transfer to monomer	$R_n^i + M_j$	$\xrightarrow{k_{fmij}}$	$P_n + R_1^j + (j-1)PDB$
Termination[b]	$R_n^i + R_m^i$	$\xrightarrow{k_{tij}}$	$P_n + P_m$
Cross-linking	$R_n^i + PDB$	$\xrightarrow{f_{PDB} k_{pi2}}$	R_{n+1}^2

[a] For consecutive reactions of the formed initiator radicals applies the initiator efficiency f_I

[b] Termination modeled by disproportionation mechanism and $k_{tij} = (k_{tii} k_{tjj})^{0.5}$

which describes the particle number density $\mathcal{F}(r,t)$ in moles as a function of the particle radius r and time t (cf. Kiparissides (2006); Vale & McKenna (2005)). The time derivative of $\mathcal{F}(r,t)$ is a function of the velocity of radial particle growth $v(r)$, nucleation with rate $\mathcal{R}_{nuc}(r)$ and coagulation among particles with rate $\mathcal{R}_{coag}(r)$. Particle nucleation therein occurs exclusively at the nucleation radius r_{nuc}, which is expressed by the Kronecker delta function $\delta_{r_{nuc},r}$. Integration of $\mathcal{F}(r,t)$ over the particle radius provides the absolute number of particles N^P,

$$N^P = V N_A \int_{r_{min}}^{r_{max}} \mathcal{F}(r,t) dr \qquad (3.2)$$

where N_A is the Avogadro constant and V the volume of the reactor content. Note that for simplicity, time dependence of variables is not stated explicitely in the following.

The free radical copolymerization reactions assumed to take place in both phases are listed in Table 3.1. Therein, the basic assumptions concerning the polymerization reactions are inherited from the process model described in Chapter 2.2.1. The species initiator I, initiator radicals $I^\bullet$, monomers M_i, radicals R^i, pendant double bonds PDB and dead polymer P refer to the respective phase (liquid phase l and gel particles g). The indices i, j correspond to the species of monomer or terminal end of radical ($i, j = 1$ for VCL and $i, j = 2$ for BIS) and n and m denote the length of the radical or polymer.

Based on the reaction mechanisms above, the terms of the pseudo-bulk model are formulated in the following. First, the formulation of particle nucleation, represented by $\mathcal{R}_{nuc}$ in Eq. (3.1), is described as well as the consecutive monomer partitioning among the liquid phase and the recently formed particles. Then, particle growth due to ongoing polymerization, represented by $v(r)$, and the average number of particles with radical entry and desorption are described. Further, the formulation for the coagulation rate $\mathcal{R}_{\text{coag}}$ in Eq. (3.1) is established, before a population balance equation for the PDB of the cross-linker is introduced to account for the cross-linking reactions in the particles. Finally, the calculation of reaction enthalpy transfer rate and properties of particle size distribution for the combination of model and experimental data are described.

3.2.1 Particle nucleation and monomer partitioning

The nucleation rate $\mathcal{R}_{\text{nuc}}$ in Eq. (3.1) describes the formation of precursor particles. These particles are formed when the dissolved oligomers precipitate due to lower solubility at high chain length. Corresponding to the homogeneous nucleation in emulsion polymerization, this is expressed by the propagation beyond a discrete critical chain length with η repeating units (cf. Hansen & Ugelstad (1978); Immanuel et al. (2002)). Hence, the nucleation rate is obtained by

$$\mathcal{R}_{\text{nuc}} = \frac{V^{\text{l}}}{V} \Sigma_{i,j=1}^{2} k_{\text{p}ij}^{\text{l}} p_i^{\text{l}} c_{\text{M}_j}^{\text{l}} c_{\text{R}_\eta}^{\text{l}}, \tag{3.3}$$

with $k_{\text{p}ij}^{\text{l}}$ as the propagation rate constant of monomers of type j with radicals with terminal end i, $c_{\text{R}_\eta}^{\text{l}}$ as the concentration of radicals with length η, p_i^{l} as the fraction of radicals with terminal end of species i, $c_{\text{M}_j}^{\text{l}}$ as concentrations of monomers of species j and V^{l} as the volume of the liquid phase. The derivation of p_i^{l} and $c_{\text{R}_\eta}^{\text{l}}$ based on the quasi-homopolymerization approach (Storti et al. (1989)) and quasi-steady state assumption is described in Appendix D.1.

The nucleation radius r_{nuc} is calculated assuming spherical precursor particles. Each precursor particle consists of one oligomer with η repeating units, each with the molar mass $M_{\text{w,unit}}$ (cf. Appendix D.1, Eq. ((D.13)), and with a residue water mass fraction w_{W}^{g}.

$$r_{\text{nuc}} = \sqrt[3]{\frac{3\eta M_{\text{w,unit}}}{(1-w_{\text{W}}^{\text{g}})4\pi\rho^{\text{g}} N_{\text{A}}}} \tag{3.4}$$

The density ρ^{g} of the microgel particle is derived from the polymer density ρ_{P} and the water density ρ_{W} under the assumption of incompressible volumes of water and polymer.

$$\rho^{\text{g}} = \frac{\rho_{\text{W}}\rho_{\text{P}}}{\rho_{\text{W}}(1-w_{\text{W}}^{\text{g}}) + \rho_{\text{P}} w_{\text{W}}^{\text{g}}} \tag{3.5}$$

The total volume of the reactor content V consists of the volumes of the two phases $V =$

$V^{\mathrm{l}} + V^{\mathrm{g}}$. The volume of the gel phase V^{g} is calculated by the integral of the volumes of the individual particles $V^{\mathrm{p}}(r)$

$$V^{\mathrm{g}} = V N_{\mathrm{A}} \int_{r_{\min}}^{r_{\max}} V^{\mathrm{p}}(r) \mathcal{F}(r,t) \mathrm{d}r, \tag{3.6}$$

where $V^{\mathrm{p}}(r)$ is calculated by $V^{\mathrm{p}}(r) = \frac{4}{3}\pi r^3$.

For smaller molecules involved in the reactions in Table 3.1, such as initiator I, monomer $\mathrm{M_1}$ and cross-linker $\mathrm{M_2}$, transport limitations are neglected and the concentrations in liquid phase and gel particles are assumed to be determined by phase equilibrium. A similar assumption is proposed by Arosio et al. (2011a). This is represented by the partition coefficient P_o.

$$P_o = \frac{c_o^{\mathrm{g}}}{c_o^{\mathrm{l}}}, \quad o \in \{\mathrm{I}, \mathrm{M_1}, \mathrm{M_2}\} \tag{3.7}$$

Hence, the balances for the overall amount of substance n and the concentrations c in the respective phases can be formulated as:

$$\frac{\mathrm{d}n_{\mathrm{I}}}{\mathrm{d}t} = -V^{\mathrm{l}} k_{\mathrm{d}} c_{\mathrm{I}}^{\mathrm{l}} - V^{\mathrm{g}} k_{\mathrm{d}} c_{\mathrm{I}}^{\mathrm{g}} \tag{3.8}$$

$$\begin{aligned} \frac{\mathrm{d}n_{\mathrm{M}_i}}{\mathrm{d}t} &= -V^{\mathrm{l}} c_{\lambda_0}^{\mathrm{l}} c_{\mathrm{M}_i}^{\mathrm{l}} \sum_{j=1}^{2} \left(k_{\mathrm{p}ji}^{\mathrm{l}} + k_{\mathrm{fm}ji}^{\mathrm{l}} \right) p_j^{\mathrm{l}} \\ &\quad - V c_{\mathrm{M}_i}^{\mathrm{g}} \sum_{j=1}^{2} \left(k_{\mathrm{p}ji}^{\mathrm{g}} + k_{\mathrm{fm}ji}^{\mathrm{g}} \right) p_j^{\mathrm{g}} \int_{r_{\min}}^{r_{\max}} \bar{n}(r) \mathcal{F}(r,t) \mathrm{d}r, \quad i = 1,2 \end{aligned} \tag{3.9}$$

$$n_o = c_o^{\mathrm{l}} V^{\mathrm{l}} + c_o^{\mathrm{g}} V^{\mathrm{g}} \quad o \in \{\mathrm{I}, \mathrm{M_1}, \mathrm{M_2}\} \tag{3.10}$$

k_{d} is the initiator decomposition rate and $k_{\mathrm{fm}ij}$ denote the chain transfer to monomer rate constants of radicals with terminal end i and monomer of species j. $c_{\lambda_0}^{\mathrm{l}}$ represents the concentration of radicals in the liquid phase (cf. Appendix D.1) and $\bar{n}(r)$ represents the average number of radicals in a particle of radius r, while p_j is the fraction of radicals of species j in the respective phase. The temperature dependence of the initiation, chain propagation and chain transfer to monomer reaction rate constants is expressed in terms of the Arrhenius equation

$$k = A \exp\left(\frac{-E_{\mathrm{A}}}{RT}\right), \tag{3.11}$$

where A is the frequency factor, E_{A} is the activation energy, R is the universal gas constant and T is the reactor temperature.

3.2.2 Particle growth

The particles continue to grow by chain propagation of the radicals inside a particle. Hence, the radial growth rate of particles with radius r results from the addition of polymer volume and water,

$$v(r) = \frac{M_{\mathrm{w,unit}}}{(1-w_{\mathrm{W}}^{\mathrm{g}})N_{\mathrm{A}}\rho^{\mathrm{g}}4r^2\pi} \sum_{i,j=1}^{2} k_{\mathrm{p}ij}^{\mathrm{g}} p_i^{\mathrm{g}} \bar{n}(r) c_{\mathrm{M}_j}^{\mathrm{g}}. \tag{3.12}$$

p_i^{g} and $c_{\mathrm{M}_j}^{\mathrm{g}}$ denote the species of terminal end of the radical and the monomer concentration in the gel phase, respectively. The average number of radicals per particle $\bar{n}(r)$ in Eq. (3.12) is derived from the Smith-Ewart theory (Nomura et al. (2005); Smith & Ewart (1948)), for which steady state is assumed

$$\mathcal{R}_{\mathrm{I}}^{\mathrm{p}}(r) + \mathcal{R}_{\mathrm{e}}^{\mathrm{p}}(r) = \mathcal{R}_{\mathrm{des}}^{\mathrm{p}}(r) + 2\mathcal{R}_{\mathrm{t}}^{\mathrm{p}}(r). \tag{3.13}$$

The rates of formed and entered radicals per particle, $\mathcal{R}_{\mathrm{I}}^{\mathrm{p}}(r)$ and $\mathcal{R}_{\mathrm{e}}^{\mathrm{p}}(r)$, equal the rates of exited and terminated radicals, $\mathcal{R}_{\mathrm{des}}^{\mathrm{p}}(r)$ and $\mathcal{R}_{\mathrm{t}}^{\mathrm{p}}(r)$, with the last two being functions of $\bar{n}(r)$. The superscript p represents that these rates refer to individual particles of size r. As a consequence of initiator partitioning, Eq. (3.13) also includes the radical formation by initiator decomposition in the particle,

$$\mathcal{R}_{\mathrm{I}}^{\mathrm{p}}(r) = 2 f_{\mathrm{I}}^{\mathrm{g}} k_{\mathrm{d}} c_{\mathrm{I}}^{\mathrm{g}} N_{\mathrm{A}} V^{\mathrm{p}}(r). \tag{3.14}$$

with initiator efficiency $f_{\mathrm{I}}^{\mathrm{g}}$. The termination rate in a particle with the pseudo-bulk assumption and for copolymerization (cf. Immanuel et al. (2002); Smith & Ewart (1948)),

$$\mathcal{R}_{\mathrm{t}}^{\mathrm{p}}(r) = \frac{n(r)^2}{N_{\mathrm{A}} V^{\mathrm{p}}(r)} \sum_{i,j=1}^{2} k_{\mathrm{t}ij}^{\mathrm{g}} p_i p_j, \tag{3.15}$$

includes the termination rate constants $k_{\mathrm{t}ij}^{\mathrm{g}}$, which is distinguished into the chemical termination contribution (=$k_{\mathrm{t}ij}^{\mathrm{l}}$) and a diffusion-controlled contribution ($k_{\mathrm{t,diff}ii}^{\mathrm{g}}$) by: ${k_{\mathrm{t}ii}^{\mathrm{g}}}^{-1} = {k_{\mathrm{t}ii}^{\mathrm{l}}}^{-1} + {k_{\mathrm{t,diff}ii}^{\mathrm{g}}}^{-1}$ (Buback et al. (1992)). The diffusion-controlled contribution is assumed to be independent of the conversion as a high polymer content ensues immediately from precipitation in the particles (cf. Chapter 2.2.1).

Radical entry and desorption The entry of oligomers from the liquid phase into a particle is described by the entry rate $\mathcal{R}_{\mathrm{e}}^{\mathrm{p}}(r)$. For $\mathcal{R}_{\mathrm{e}}^{\mathrm{p}}(r)$, the diffusion controlled radical capture model by Smith & Ewart (1948) is employed,

$$\mathcal{R}_{\mathrm{e}}^{\mathrm{p}}(r)=\frac{f_{\mathrm{e}}4\pi rN_{\mathrm{A}}c_{\lambda_0}^{\mathrm{l}}}{\sqrt{l_{\mathrm{avg}}^{\mathrm{l}}}}\sum_{i=1}^{2}D_{\mathrm{W}_i}p_i^{\mathrm{l}} \tag{3.16}$$

which assumes that diffusion of oligomers in the liquid phase determines radical entry. The diffusion coefficient of monomers in the liquid phase, D_{W_i}, is adjusted by the average oligomer length $l_{\mathrm{avg}}^{\mathrm{l}}$ (cf. Appendix D.1, Eq. (D.3)) in the liquid phase to describe the diffusion coefficient of oligomers. f_{e} is an efficiency factor to compensate the overprediction of the radical capture model (Nomura et al. (2005)).

For the radical desorption rate $\mathcal{R}_{\mathrm{des}}^{\mathrm{p}}(r)$ from a single particle, a simple form of the radical desorption model proposed by Harada et al. (1971) is applied

$$\mathcal{R}_{\mathrm{des}i}^{\mathrm{p}}(r)=\bar{n}(r)c_{\mathrm{M}_i}^{\mathrm{g}}\sum_{j=1}^{2}k_{\mathrm{fm}ji}^{\mathrm{g}}p_j\frac{k_{\mathrm{des}i}(r)}{k_{\mathrm{des}i}(r)+\sum_{j=1}^{2}k_{\mathrm{p}ij}^{\mathrm{g}}c_{\mathrm{M}_j}^{\mathrm{g}}}. \tag{3.17}$$

Therein, only monomeric radicals formed by chain transfer to monomer can desorb from the particles. Desorption, described by equilibrium radical desorption coefficient $k_{\mathrm{des}i}$ for the monomeric radicals of species i, competes with chain propagation, which is expressed in terms of the event probability. $k_{\mathrm{des}i}$ depends on the diffusion coefficients D_{W_i} and D_{P} in water and polymer, respectively, and the partition coefficients P_{M_i} according to the relation established by Harada et al. (1971):

$$k_{\mathrm{des}i}(r)=\frac{3D_{\mathrm{W}i}D_{\mathrm{P}}}{r^2(D_{\mathrm{W}_i}+D_{\mathrm{P}}P_{\mathrm{M}_i})}. \tag{3.18}$$

3.2.3 Particle coagulation

The particle coagulation rate $\mathcal{R}_{\mathrm{coag}}(r)$ combines formation and depletion of particles with radius r. Depletion describes the loss of particles due to their aggregation with other particles. Formation accounts for the gain of larger particles due to the aggregation of smaller particles. The coagulation rate is employed as described by Immanuel & Doyle III (2003).

$$\begin{aligned}\mathcal{R}_{\mathrm{coag}}(r)=&-\int_{r_{\min}}^{r_{\max}}\beta(r,r')\mathcal{F}(r,t)\mathcal{F}(r',t)\mathrm{d}r'\\&+\int_{r_{\min}}^{\frac{r}{\sqrt[3]{2}}}\beta(r',r'')\mathcal{F}(r',t)\mathcal{F}(r'',t)\frac{r^2}{(r^3-r'^3)^{\frac{2}{3}}}\mathrm{d}r'\end{aligned} \tag{3.19}$$

The radii of the aggregating particles are therein related by the combined volume, which equals the volume of the particle with radius r: $r'^3 + r''^3 = r^3$. The size-dependent coagulation kernel $\beta(r,\tilde{r})$ is calculated employing a semi-empirical approach. The simplification distinguishes among unstable and stable particles with respective radii below and above a stable particle radius r_{stable} (Araújo et al. (2001); Vale & McKenna (2009)). Unstable particles can coagulate with unstable particles as well as stable particles, whereas stable particles can aggregate with unstable particles but not with other stable particles. This is described by an adaptation of the Fuchs modification of the Smucholski equation as used by Vale & McKenna (2009):

$$\beta(r,\tilde{r}) = 0.5(1 - \tanh(r^* - r_{\text{stable}}))\frac{2k_B T N_A}{3\mu^l W}\left(2 + \frac{r}{\tilde{r}} + \frac{\tilde{r}}{r}\right). \tag{3.20}$$

r^* denotes the smaller particle radius of the particles involved, r or r'. k_B, μ^l and W represent the Boltzmann constant, the viscosity of the liquid phase and the Fuchs stability ratio, respectively. The distinction among unstable and stable particles ensures that particle aggregation will come to a hold, which is essential for a monodisperse particle size distribution.

3.2.4 Pendant double bond density

PDB are tied to a specific polymer particle by reaction. Hence, the PDB are balanced in dependence of the particle radius r, comparable to the particle density $\mathcal{F}(r,t)$. The distribution of the pseudo-species PDB, $\mathcal{F}_{\text{PDB}}(r,t)$, is formulated in terms of its moles in particles of radius r. Hence, it can be interpreted as the product of $\mathcal{F}(r,t)$, V and the average number of PDB per particle with radius r.

$$\begin{aligned}
\left.\frac{\partial \mathcal{F}_{\text{PDB}}}{\partial t}\right|_{r,t} + \left.\frac{\partial \mathcal{F}_{\text{PDB}} \nu}{\partial r}\right|_{r,t} &= \delta_{r_{\text{nuc}},r} p_2^l \mathcal{R}_{\text{nuc}} V^l \eta \\
&+ V\mathcal{F}(r,t)\bar{n}(r) c_{\text{M}_2}^{\text{g}} \sum_{i=1}^{2}\left(k_{\text{p}i2}^{\text{g}} + k_{\text{fm}i2}^{\text{g}}\right) p_i^{\text{g}} - \mathcal{F}_{\text{PDB}}(r,t)\frac{\bar{n}(r)}{N_A V^{\text{p}}}\sum_{i=1}^{2} f_{\text{PDB}}^{\text{g}} k_{\text{p}i2}^{\text{g}} p_i^{\text{g}} \\
&+ \mathcal{F}(r,t)V\left(\mathcal{R}_{\text{e}}^{\text{p}}(r) p_2^l l_{\text{avg}}^l - \mathcal{R}_{\text{des2}}^{\text{p}}(r)\right) - \int_{r_{\min}}^{r_{\max}} \beta(r,r')\mathcal{F}_{\text{PDB}}(r,t)\mathcal{F}(r',t)\,\mathrm{d}r' \\
&+ \int_{r_{\min}}^{\frac{r}{\sqrt[3]{2}}} \beta(r',r'')\left(\mathcal{F}_{\text{PDB}}(r',t)\mathcal{F}(r'',t) + \mathcal{F}_{\text{PDB}}(r'',t)\mathcal{F}(r',t)\right)\frac{r^2}{(r^3 - r'^3)^{\frac{2}{3}}}\mathrm{d}r'
\end{aligned} \tag{3.21}$$

Like the particle number density, $\mathcal{F}_{\text{PDB}}(r,t)$ depends on precipitation, growth and coagulation. In addition, $\mathcal{F}_{\text{PDB}}(r,t)$ is also affected by the formation of new PDB in a particle by reactions of cross-linker, cross-linking itself with cross-linking efficiency $f_{\text{PDB}}^{\text{g}}$ as well as radical ab- and desorption. For the gain of PDB by precipitation and absorption, the number of entering

PDB is calculated from the length of the entering polymer chains, η and $l^{\mathrm{l}}_{\mathrm{avg}}$, respectively, and the fraction of incorporated cross-linker, which is approximated by the fraction of radicals with terminal end of type cross-linker p^{l}_{2}. For loss of PDB by desorption, the desorption rate for radicals of type cross-linker is employed and, since only radicals with one repeating unit can desorb, it is not required to consider the chain length here.
With the cross-linking reaction, the terminal end of the reacting oligomer changes to a radical end of species cross-linker. Hence, the Mayo-Lewis equation to calculate the fraction p^{g}_{i} of radicals with species i is extended by the pseudo-species PDB

$$p^{\mathrm{g}}_{i} \sum_{j=1}^{2} \left(k^{\mathrm{g}}_{\mathrm{p}ij} c^{\mathrm{g}}_{\mathrm{M}_j} + f^{\mathrm{g}}_{\mathrm{PDB}} k^{\mathrm{g}}_{\mathrm{p}i2} c^{\mathrm{g}}_{\mathrm{PDB}} \right) = \sum_{j=1}^{2} p^{\mathrm{g}}_{j} k^{\mathrm{g}}_{\mathrm{p}ji} c^{\mathrm{g}}_{\mathrm{M}_i}, \quad i = 1 \tag{3.22}$$

$$\sum_{i=1}^{2} p^{\mathrm{g}}_{i} = 1 \tag{3.23}$$

with the average PDB concentration $c^{\mathrm{g}}_{\mathrm{PDB}}$

$$c^{\mathrm{g}}_{\mathrm{PDB}} = \frac{1}{V^{\mathrm{g}}} \int_{r_{\mathrm{min}}}^{r_{\mathrm{max}}} \mathcal{F}_{\mathrm{PDB}}(r,t)\mathrm{d}r \tag{3.24}$$

3.2.5 Particle characterization

The particle size distribution is characterized by the average microgel radius r_{h} and the standard deviation of the particle size distribution σ. The Polydispersity Index *PDI*, as defined for Dynamic Light Scattering (DLS) measurements, is a measure for the width of the particle size distribution. [1]

$$r_{\mathrm{h}} = \frac{\int_{r_{\mathrm{min}}}^{r_{\mathrm{max}}} r\mathcal{F}(r,t)\mathrm{d}r}{\int_{r_{\mathrm{min}}}^{r_{\mathrm{max}}} \mathcal{F}(r,t)\mathrm{d}r} \tag{3.25}$$

$$\sigma = \sqrt{\frac{\int_{r_{\mathrm{min}}}^{r_{\mathrm{max}}} (r - r_{\mathrm{h}})^2 \mathcal{F}(r,t)\mathrm{d}r}{\int_{r_{\mathrm{min}}}^{r_{\mathrm{max}}} \mathcal{F}(r,t)\mathrm{d}r}} \tag{3.26}$$

$$PDI = \left(\frac{\sigma}{r_{\mathrm{h}}} \right)^2 \tag{3.27}$$

[1] nanoComposix - Guidelines for DLS Measurement and Analysis, `www.nanocomposix.com/pages/knowledge-base`, Accessed: 22.11.2018

3.2.6 Reaction enthalpy transfer rate

The reaction enthalpy transfer rate Σ_R represents the enthalpy released when double bonds react in a propagation reaction. Considering the propagation reactions in both the liquid phase and the gel particles, and propagation reactions with pendant double bonds, Σ_R is calculated according to

$$\begin{aligned}\Sigma_\mathrm{R} = &- V^\mathrm{l} c^\mathrm{l}_{\lambda_0} \sum_{i,j=1}^{2} \left(\Delta H_{\mathrm{R}ij} k^\mathrm{l}_{\mathrm{p}ij} p^\mathrm{l}_i c^\mathrm{l}_{\mathrm{M}_j} \right) \\ &- V \sum_{i,j=1}^{2} \Delta H_{\mathrm{R}ij} k^\mathrm{g}_{\mathrm{p}ij} p^\mathrm{g}_i c^\mathrm{g}_{\mathrm{M}_j} \int_{r_\mathrm{min}}^{r_\mathrm{max}} \bar{n}(r) \mathcal{F}(r,t) \mathrm{d}r \\ &- \sum_{i=1}^{2} \Delta H_{\mathrm{R}i2} \frac{f^\mathrm{g}_\mathrm{PDB} k^\mathrm{g}_{\mathrm{p}i2} p^\mathrm{g}_i}{N_\mathrm{A}} \int_{r_\mathrm{min}}^{r_\mathrm{max}} \frac{\bar{n}(r) \mathcal{F}_\mathrm{PDB}(r,t)}{V^\mathrm{p}(r)} \mathrm{d}r. \end{aligned} \tag{3.28}$$

$\Delta H_{\mathrm{R}ij}$ represents the enthalpy of the specific propagation reaction of a radical with terminal end j with monomer i. As described in Chapter 2.2.2, the enthalpy transfer rate links the model with experimental data from reaction calorimetry.

3.3 Experimental data

The recipes for the microgel syntheses are listed in Table 3.2. All reaction components except initiator are initially dissolved in the solvent, and the reaction mixture is heated under stirring to the reaction temperature. The reactor is purged for 30 min with nitrogen. When steady state is obtained, the initiator is added and the synthesis runs for 1 h. After synthesis, the reactor content is cooled to room temperature and the microgels are dialysed. The specifications for monomer and initiator concentration as well as temperature for the reference experiment are defined according to the common and well-tested recipe for PVCL-based microgels by precipitation polymerization (e.g. Balaceanu et al. (2011); Schneider et al. (2014); Wolff et al. (2018)), and corresponds to the experiment with 2.5 mol-% BIS in Chapter 2. The recipe variations include changes of temperature as well as initial initiator or cross-linker concentrations, modified one at a time.

From the measurements of the Mettler Toledo reaction calorimeter RTcal in isothermal control mode, the reaction enthalpy transfer rate Σ_R is calculated with the energy balance for the isothermal batch reactor (cf. Chapter 2.2.2). The experimental data of reaction calorimetry for the reference experiment and the variation of the initial cross-linker concentrations is the same as used for the terpolymerization case study (cf. Appendix C).

Table 3.2: Recipes for the reference microgel synthesis and the investigated experiment variations of different temperatures, initial initiator or cross-linker concentrations. Information is given as initial conditions for Eqs. (3.8)-(3.9). The labels refer to the initiator or cross-linker to monomer ratios in the respective experiment variations.

Label	$n_{M_1}(t=0\,\text{s})$ mol	$n_{M_2}(t=0\,\text{s})$ mol	$n_I(t=0\,\text{s})$ mol	T K
reference	0.0318	$7.78\cdot10^{-4}$	$3.69\cdot10^{-4}$	343
333 K 353 K	0.0318	$7.78\cdot10^{-4}$	$3.69\cdot10^{-4}$	333, 353
0.6 mol-% I 2.4 mol-% I	0.0318	$7.78\cdot10^{-4}$	$1.84\cdot10^{-4}$, $7.37\cdot10^{-4}$	343
1.2 mol-% BIS 5.0 mol-% BIS	0.0318	$3.89\cdot10^{-4}$, $1.56\cdot10^{-3}$	$3.69\cdot10^{-4}$	343

All experiments are performed in 0.3 L water as solvent with 0.0443 g CTAB as stabilizer.

The final microgel size in collapsed state is determined by Dynamic Light Scattering at 323 K. It has been reported previously that dialysis does not affect the measured microgel radius in collapsed state (Imaz & Forcada (2008b)), hence microgel characterization is performed after dialysis. For the reference experiment, the microgel radius is determined by the mean of 5 experiment repetitions with a standard deviation $\sigma_{r_h} = 3.8\,\text{nm}$. The microgel radii for the experiment variations are determined from single experiments.

3.4 Simulation and parameter estimation

The model is implemented in gProms Model Builder, version 5.0.1 [1]. The discretization of the partial differential equations (PDE) in Eqs. (3.1) and (3.21) are performed by the gProms intrinsic first-order Backwards Finite Differences Method. To reduce the error of numerical diffusion of the selected solution method, a fine discretization of 250 intervals is chosen. For dynamic integration, the gProms intrinsic DASOLV solver is selected, which is based on a variable time step and variable order Backward Differentiation Formula. The employed parameter values for rate constants and reaction enthalpies for propagation and chain transfer to monomer are listed in Table A.2 and Table A.3 in the Appendix. Parameter values and equations for initiator decomposition, diffusion coefficients, partition coefficients, critical chain

[1] Process Systems Enterprise, gProms, www.psenterprise.com/gproms 1997-2020

length, water density and viscosity are provided in Table A.4. Parameter estimation of the adjustable parameters is conducted with the maximum likelihood method with the same solution parameters as selected for dynamic integration.

Six adjustable parameters (W, r_{stable}, f_e, D_P and $k^g_{t,\text{diff}ii}$) are determined by parameter estimation. The termination rate constants $k^g_{t,\text{diff}ii}$ have been previously estimated with the two-phase model (cf. Appendix C for the two-phase model with the simplification described above). Unpublished simulation studies with the previous model have shown that temperature-independent termination rate constants provide sufficiently accurate predictions in the considered temperature range. Hence, a temperature dependence of the termination rate constants k^l_{tii} and $k^g_{t,\text{diff}ii}$ is neglected. Nonetheless, considering additional mass transfer among the phases (radical entry and desorption) and distinct particle volumes instead of a continuous gel phase implies a change of the average radical concentration and consequently demands an adjustment of the termination rate constant $k^g_{t,\text{diff}ii}$. The previously determined values for $k^g_{t,\text{diff}ii}$ are used as initial guesses for parameter estimation.

The parameter values are estimated based on the reference experiment (cf. Table 3.2). Experimental data for the parameter estimation includes the reaction enthalpy transfer rate Σ_R and the final particle radius r_h as well as the condition for a monodisperse PSD ($PDI = 0.01$). For parameter estimation, the interval of 700 s after initiation ($t = 0$ s) is used, because reaction calorimetry shows only within this time frame significant dynamic behavior. The number of Σ_R measurements, which are available in 2 s intervals, outweighs the end-point measurements of r_h and PDI. Since the gProms intrinsic parameter estimation does not allow to weigh the experimental data, the measurements of r_h and PDI are duplicated over the interval of $t = 600 - 700$ s. Within this interval, calorimetry measurements are in steady state. The introduction of artificial measurement points compensates the high number of data points for Σ_R compared to few data points for r_h and PDI and will ensure that the measurements of the particle radius are respected.

3.5 Results and discussion

In the following, the simulation of the fitted model is compared to the experimental data. First, the results of the parameter estimation are evaluated by a comparison of simulation with the fitted model and experimental data. This comparison includes the reaction enthalpy transfer rate (Eq. (3.28)), the average particle radius (Eq. (3.25)) and the polydispersity (Eq. (3.27)). Afterwards, the fitted model is employed to predict particle growth for the variation of reaction conditions and the predictions are compared to the corresponding experimental data.

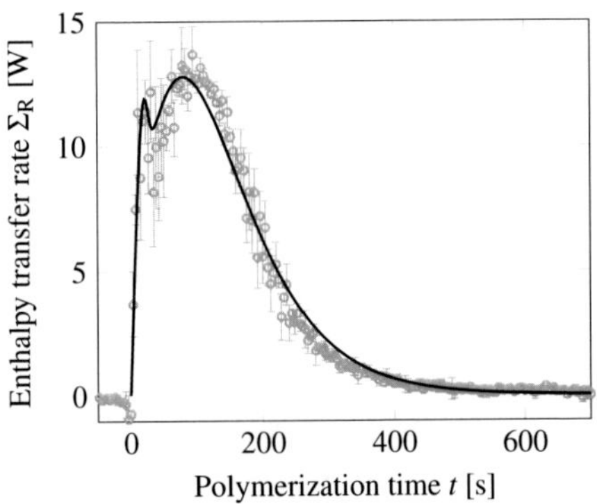

Figure 3.2: Comparison of experimental data (cf. Appendix C) and simulation with fitted model for a polymerization of the reference experiment (cf. Table 3.2). The error bars denote the standard deviations of measurements of three experimental repetitions. For parameter estimation, a constant mean variance model is applied ($\sigma = 0.32$ W).

3.5.1 Simulation of reference experiment

Figure 3.2 shows the comparison of the simulation and the experimental data in terms of Σ_R over the polymerization time t. The experimental data is used as published previously. $t = 0$ s represents the time of initiation and the beginning of the simulation. Immediately after initiation, Σ_R increases rapidly to a first maximum after approximately $t = 25$ s, caused by the fast cross propagation reaction of VCL and BIS (cf. Chapter 2.2.5). Then, the enthalpy transfer rate decreases due to consumed cross-linker, before it increases again due to the polymerization of the remaining VCL. The simulation with the fitted model agrees with the experimental data within the accuracy of the standard deviation. In conclusion, the pseudo-bulk model is equally suited to fit reaction enthalpy measurements as the two-phase model presented previously.

Besides the reaction calorimetry, the simulation satisfies the experimental values for the mean particle radius. Figure 3.3 shows the predicted growth of the average particle radius over the polymerization time. The average particle radius increases rapidly immediately after initiation and after approx. $t = 400$ s, the microgels have obtained their final particle size. At the end of the simulation, the predicted average particle radius shows a very good agreement with the particle radius from DLS and the deviation is below the experimental error. For further insight, the prediction of the mean particle radius is validated with data from *in situ*

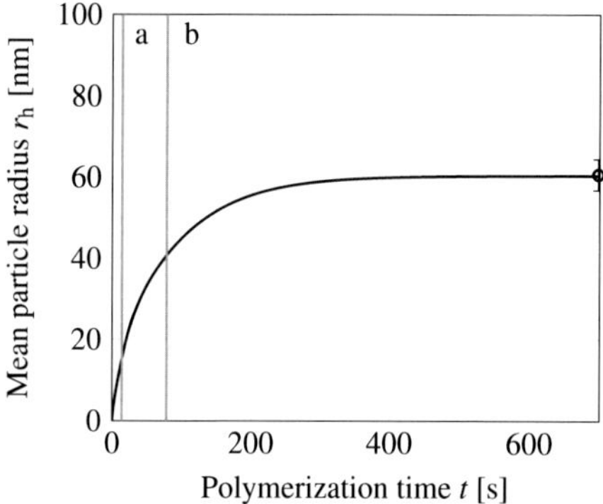

Figure 3.3: Predicted mean particle radius over polymerization time and measured final radius (DLS) for the reference experiment (cf. Table 3.2). The deviation is below the standard deviation of the measurements. Vertical lines illustrate the predicted ends of (a) nucleation and (b) coagulation.

DLS, which is provided in the Appendix D.2.

Finally, the simulation needs to satisfy the condition of a monodisperse PSD. Figure 3.4 shows the PSD in terms of the particle number density $\mathcal{F}(r,t)$ over the particle radius at different stages of the synthesis. At early stages ($t = 10-50\,\text{s}$), coagulation has a significant effect on the particle number density, as the area under curves, proportional to the particle number, decreases rapidly. In the further progress ($t = 100-400\,\text{s}$), coagulation has come to a hold and radical and monomer absorption merely shift the mean of $\mathcal{F}(r,t)$ to higher radii, while the particle number, represented by the area under the curves, remains unaffected. At late stages ($t = 400\,\text{s}, 700\,\text{s}$), particle growth has come to an end as the particle number densities cannot be distinguished. Throughout the synthesis, the predicted particle density shows a monodisperse PSD is obtained despite the inhomogeneous consumption of cross-linker and monomer in Figure 3.2, since small particles are absorbed instead of leading to a secondary nucleation. Based on *in situ* small-angle neutron scattering (SANS) measurements of the synthesis of PNIPAM-based microgels, Virtanen et al. (2019) recently concluded a comparable formation mechanisms of a short initial nucleation phase and homogeneous growth of microgels thereafter.

For comparison of the final particle size distribution, the intensity measurement from DLS for an individual sample is provided. The intensity measurement reveals a narrow distribution around the mean particle radius. The comparison shows that the simulated $\mathcal{F}(r,t)$ provides a

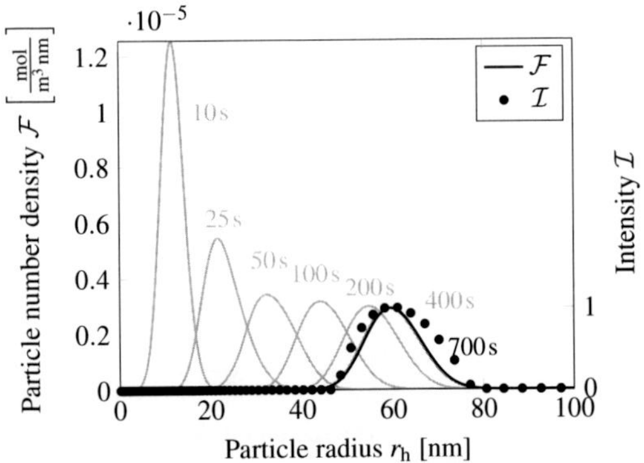

Figure 3.4: Prediction of particle size distribution for the reference experiment (cf. Table 3.2) at different stages of the process. For comparison of the final particle size distribution, intensity measurements from DLS for an individual sample (323 K) are included. For parameter estimation, a $PDI = 0.01$ for the measured r_h was used.

Table 3.3: Estimated parameter values for pseudo-bulk model. Additional remarks regarding their interpretation or context of literature values are provided in Appendix D.3.

Parameter	Symbol	Unit	Parameter value
Radical entry efficiency	f_e		0.534
Diffusion coefficient	D_P	m^2s^{-1}	$10^{-16.65}$
Diffusion limited contribution of	$k^g_{t,diff11}$	$m^3(mol\ s)^{-1}$	$10^{-0.90}$
termination rate constants	$k^g_{t,diff22}$	$m^3(mol\ s)^{-1}$	$10^{1.64}$
Stable particle radius	r_{stable}	nm	26.27
Fuchs stability ratio	W		$10^{4.44}$

good prediction of the monodisperse particle size distribution of the final microgels.

The estimated parameter values, listed in Table 3.3, show the general trend of the impacts of the individual growth mechanisms. Three central aspects should be highlighted here: (1) The radical entry efficiency is high, compared to values proposed from literature. It is probably exaggerated due to the inaccuracy of pseudo-bulk models to describe nucleation. Nonetheless, there remains an uncertainty regarding the values from literature since these are determined for emulsion polymerization and might not apply for precipitation polymerization. (2) The diffusion-controlled contributions of the termination rate constants estimated with the pseudo-bulk model provide the same trends as those previously estimated with the two-phase model, but cannot be transferred directly under the given assumptions. The larger value of $k^{g}_{t,diff22}$ affects only the beginning of the process before the cross-linker is fully consumed. Hence, the diffusion-controlled contribution effectively decreases in value over time (and conversion), which could represent an increasing diffusion limitation from gelation in the particles. (3) The results for the stable particle radius and Fuchs stability ratio suggest that coagulation has a significant impact on particle growth and is not limited to precursor particles.

It should also be noted that despite the effort to limit the number of adjustable parameters for the investigation of the reaction system, the estimated parameter values reveal in parts significant correlations. This concerns the values of the parameters W, r_{stable} and f_e, but also the combination of parameter values for D_P and $k^{g}_{t,diff11}$. These high correlations indicate that parameter values are not identifiable based on the available experimental data. For instance, both coagulation parameters r_{stable} and W have a significant effect on the particle size distribution (cf. Appendix D.4). Higher r_{stable} lead to larger particle radii since larger particles coagulate, broadening the particle size distributions in terms of higher *PDI*. Decreasing W increases the coagulation kernel and therefore accelerates particle aggregation, leading to larger particle radii and more narrow *PDI*. Hence, both an increased r_{stable} and a decreased W result in larger particle radii. The difference in the final predicted *PDI* can be small, especially in comparison to the experimental error. However, more distinct differences in the *PDI* can be observed at earlier polymerization times. Hence, when particle size distribution measurements at an early stage of polymerization are provided, correlations among the coagulation parameters can be reduced, and parameter estimation results improved. Also, further model simplification can reduce or eliminate correlations, e.g., neglecting radical desorption due to its low impact, which eliminates the parameter D_P. Thus, the estimated parameter values need to be treated carefully in terms of model predictions, as the study of the predictive capabilities in the following will show.

3.5.2 Prediction of experiment variations

To evaluate the predictive capabilities of the fitted model, variations of the microgel synthesis are simulated and compared to experimental data. The comparison is performed in terms of reaction enthalpy transfer rates and final microgel size for variations of reaction temperature, initial initiator and cross-linker concentration.

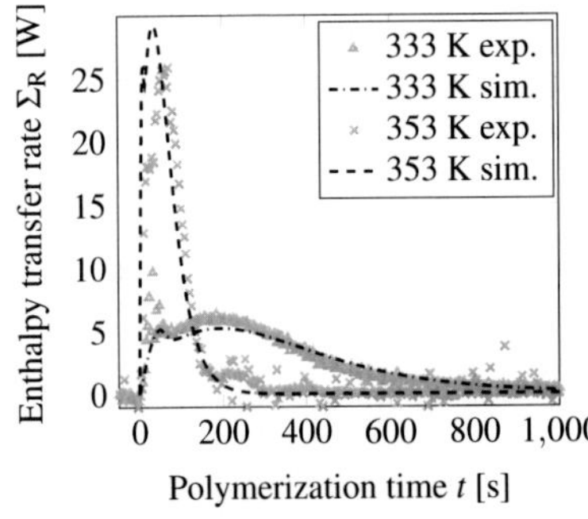

Figure 3.5: Comparison of experimental and predicted reaction enthalpy transfer rates for different reaction temperatures. Experimental data has not been published previously. For improved readability, measurement data is shown only for 4 s intervals.

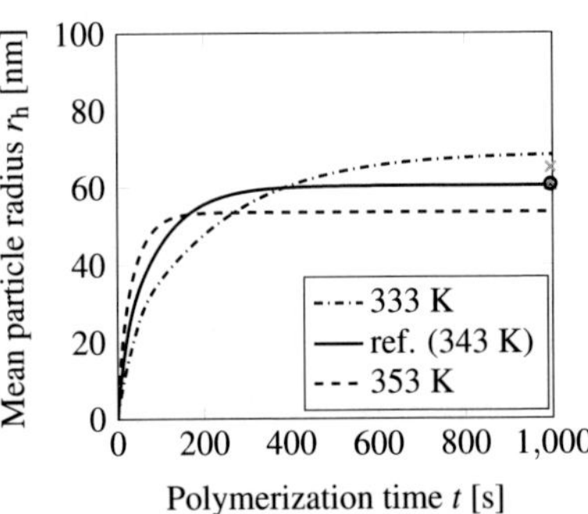

Figure 3.6: Comparison of predicted radii and measured final radii (DLS) for different reaction temperatures (cf. Table 3.2). Markers illustrate the corresponding measured particle radii at the end of synthesis (△: 333 K, ○: ref. (343 K), ×: 353 K).

3.5.2.1 Variation of reaction temperature

The first simulation addresses the impact of the reaction temperature on the overall polymerization time as well as the final particle size.

The comparison of predicted and measured Σ_R for 333 K and 353 K is depicted in Figure 3.5. The simulation for both temperatures predicts that an increase of the reaction temperature facilitates the overall polymerization. For 353 K, the maximal Σ_R is higher and it approaches zero after approx. 300 s indicating the end of synthesis, while the simulation for 333 K predicts a duration of almost 1000 s for the synthesis.

The experimental data for both temperatures confirm the predictions. Comparing accuracy of the predictions, the simulation for 333 K matches the experimental data better, while the prediction for 353 K exceeds the experimental data around its maximum value. Also, the simulation does not predict the secondary increase in measured Σ_R for T = 353 K (t = 200-300 s), which probably results from a delayed heat transfer due to polymer film on the reactor surface (Virtanen et al. (2016)). Nonetheless, in terms of Σ_R, the pseudo-bulk model provides a good prediction of the temperature dependence of the process.

Figure 3.6 shows the corresponding predicted mean particle radii over the progress of polymerization. The trend reveals that with increasing T, microgels initially grow faster but obtain a lower final particle size. At the early stage, more particles are formed by nucleation due to higher initiator decomposition rate and lower η. Faster chain propagation, radical entry and coagulation lead to a high number of particles obtaining the stable particle radius. Beyond the stable particle radius, coagulation is limited and the high number of particles competes for absorption of smaller particles and monomers. In contrast to the predicted particle radii, the DLS measurements of microgels synthesized at different temperatures show no significant trend and deviations are within the experimental error of the reference experiment. There are several explanations for the differences among prediction and experimental data: First, the temperature could affect the coagulation of the particles, which would not be covered by the empirical modeling approach. Further, microgels synthesized at different temperatures might have different densities, while in this contribution, a constant density is assumed. Also, the measurement error needs to be taken into account, as DLS measurements are obtained from singular synthesis and a temperature dependence has been observed previously in literature for PVCL- (Imaz & Forcada (2008b)) and more thoroughly for PNIPAM-based microgels (Gao & Frisken (2003); Virtanen & Richtering (2014)). In conclusion, the predictions indicate a significant dependence of particle size on temperature. However, a confirmation or disprove of the goodness of the prediction would require further experimental investigation.

3.5.2.2 Variation of initial initiator concentration

As illustrated in Figure 3.7, the prediction of Σ_R increases and shifts to earlier polymerization times with increasing initiator concentrations. The initiator concentration affects the overall polymerization in a similar manner to reaction temperature, as initiator decomposes faster with increasing temperature. But unlike the temperature, the initiator concentration does not affect the propagation rate constant and in consequence, the maximal Σ_R for 2.4 mol-% I is lower than for 353 K. Again, the predicted Σ_R is in good agreement with the experimental data without further parameter adjustment, especially for 0.6 mol-% I.

The predicted particle growth for different initial initiator concentrations is depicted in Figure 3.8. Based on the estimated parameter values, the largest final particle radii are obtained for the lowest initial initiator concentration; following the same argumentation as for temperature dependence that fewer particles are formed in the beginning of the process. However, DLS measurements show the opposite behavior. As previously observed for PVCL- and PNIPAM-based microgels, an increase of the initial initiator concentration leads to an increase of the final microgel size (Gao & Frisken (2003); Imaz & Forcada (2008b); Virtanen & Richtering (2014)). Hence, based entirely on the reaction rates, the final particle size dependence on the initiator concentration cannot be predicted. An adjustment of the coagulation parameters, such as a decrease of W and/or an increase of r_{stable}, appears to be the most promising approach. However, up to this point there are too many unknown parameters and factors for a

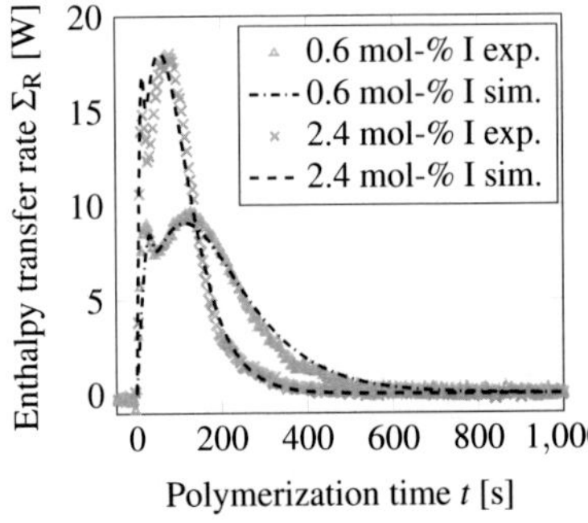

Figure 3.7: Comparison of experimental and predicted reaction enthalpy transfer rates for different initial initiator concentrations. Experimental data has not been published previoulsy. For improved readability, measurement data is shown only for 4 s intervals.

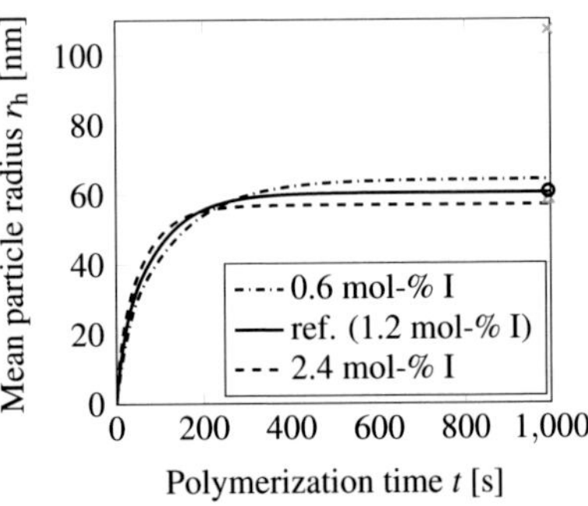

Figure 3.8: Comparison of predicted particle growth and measured final radii (DLS) for different initial initiator concentrations (cf. Table 3.2). Markers illustrate the corresponding measured particle radii at the end of synthesis (△: 0.6 mol-% I, ○: ref. (1.2 mol-% I), ×: 2.4 mol-% I).

mechanistic modeling approach (e.g., DLVO theory) and too few experimental data points for an empirical description.

3.5.2.3 Variation of initial cross-linker concentration

The predictions in Figure 3.9 show a significant impact of the initial cross-linker concentration on the Σ_R profile. The fast cross-propagation of VCL and BIS, which is discussed above in the context of Figure 3.2, amplifies with increasing the initial cross-linker concentration. This results in a rapid increase of the first peak in the predicted Σ_R profile for 5 mol-% BIS, while for 1.2 mol-% BIS, it almost disappears. The comparison to Σ_R from measurements shows that the predictive qualities of the model for overall polymerization progress in terms of the reaction enthalpy transfer rate are good, even for changing monomer/cross-linker ratios.

Though reaction rates represented by Σ_R differ significantly, the predicted particle growth for different cross-linker concentrations in Figure 3.10 differ only insignificantly. The differences in the final predicted particle radii are within the order of magnitude of standard deviation of the reference. In contrast to the variations of temperature and initial initiator concentration, the progress of the particle growth differ insignificantly in the beginning, leading to the prediction of similarly sized final microgels. For 1.2 mol-% BIS, the largest final particle size is

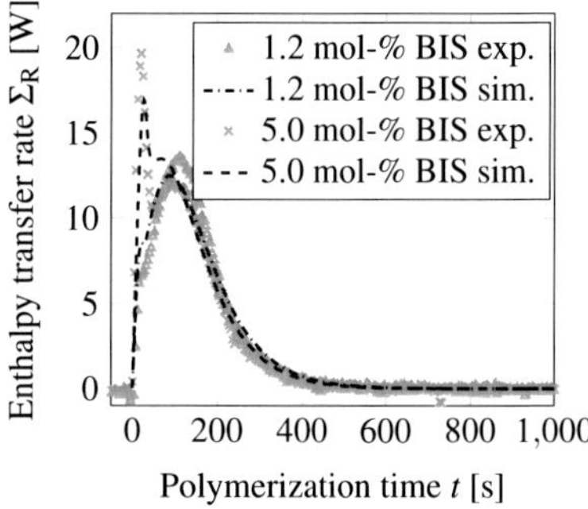

Figure 3.9: Comparison of experimental (cf. Appendix C) and predicted reaction enthalpy transfer rates for different initial cross-linker concentrations. For improved readability, measurement data is shown only for 4 s intervals.

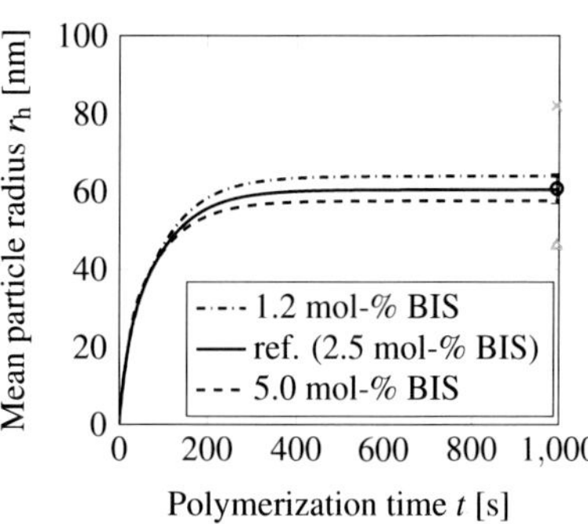

Figure 3.10: Comparison of predicted particle growth and measured final radii (DLS) for different cross-linker concentrations (cf. Table 3.2). Markers illustrate the measured particle radii at the end of synthesis (△: 1.2 mol-% BIS, ○: ref. (2.5 mol-% BIS), ×: 5.0 mol-% BIS).

predicted. This prediction is the opposite trend to the microgel sizes determined by DLS measurements. The measurements show the differences among the microgel radii are clearly larger than the standard deviation and further, larger microgels are obtained for higher initial cross-linker concentrations. Again, this leads to the conclusion that reaction rates alone cannot be employed to explain the dependence of microgel size on cross-linker concentration. An extension of the coagulation behavior, such as an increasing r_{stable} or preferably a decrease of W as function of increasing cross-linker concentration in the particles, is recommended. In addition, microgels with higher cross-linker concentration reveal higher densities due to higher internal cross-linking (Balaceanu et al. (2011); Schneider et al. (2014)). Higher microgel densities would facilitate the trend that smaller particles are obtained for higher cross-linker concentrations. These higher densities result in higher polymer fractions in the particles, which will also impact the reaction rate constants in the gel phase (Kröger et al. (2017)) and are accompanied by a higher diffusion limitation of the termination reaction. Thus, for a predictive model that captures the impact of cross-linker on the final particle size, quantitative reliable experimental data of the polymer fraction in collapsed state is required.

3.6 Conclusion

In this chapter, a pseudo-bulk model was adapted to describe the formation and growth of cross-linked PVCL-based microgels in precipitation polymerization. The purpose of this investigation was to obtain a mechanistic model that cohesively describes the most important quantities of microgel synthesis: the overall polymerization progress as well as the final microgel size with a narrow size distribution. The model comprises common formulations for all growth mechanisms in microgel synthesis: Homogeneous nucleation describes the precipitation of precursor particles and growth is described as the result of absorption of monomers and radicals as well as coagulation of the growing particles.

Based on experimental data from reaction calorimetry and DLS measurements for the final microgel size, the six adjustable parameters of the model are estimated. The comparison of simulation with the fitted model and experimental data shows that the pseudo-bulk model can be fitted to describe the characteristics of a specific synthesis process. The simulation of reaction enthalpy transfer rate agrees with the experimental data within the order of magnitude of the experimental error, while the requirements of the final microgel size are met.

The estimated parameter values suggest the impact of individual growth mechanisms. High radical entry efficiency and low effective termination rate constants in the gel phase express that the microgels are the preferred reaction locus. The stable particle radius, which is significantly larger than the nucleation radius, suggests that not only precursor particles are absorbed by larger particles but that coagulation has a significant impact on microgel growth and their size distribution.

These contributions result in the following microgel formation process: After initiation, microgels are formed in a short nucleation phase only. The growing microgels coagulate until stable, while also growing by polymerization with absorbed monomers and oligomers. Growth continues by absorption of monomers and oligomers and ongoing polymerization in the microgels until the monomer is fully consumed. The short nucleation phase and the following coagulation facilitate the narrow size distribution. This conclusion also agrees with other hypothesis based on experimental data for PNIPAM-based microgels (Virtanen et al. (2019).

Testing the predictive capabilities of the fitted model for variations of the experimental recipes demonstrates its potential and current limitations. For variations of temperature, initial initiator and cross-linker concentration, good predictions of the reaction enthalpy transfer rates were obtained. But predictions of the final microgel radii show the shortcomings and hence requirement of further research in both modeling and experimental investigation. While predictions show the largest impact on final microgel size by variation of temperature, measurements show no significant impact. In contrast, for the cross-linker concentrations, experiments show significant differences in particle radii, while simulations predict similar microgel sizes. The deviations of predictions and measurements might indicate the need for model adjustments. However, an advanced and validated model will require more quantitative information on the particles properties, such as density or water content as well as accurate measurements of microgel size throughout the synthesis process. Nonetheless, the good results for the simulation of the standard synthesis and predictions of reaction enthalpy show that the pseudo-bulk model bears the opportunity to describe the precipitation polymerization of PVCL-based microgels.

Chapter 4

Hybrid kinetic Monte Carlo model for internal microgel structure

The internal microgel structure determines the swelling behaviour and therefore has a crucial influence on microgel properties. So far, a model for the conversion to describe the polymerization progress in Chapter 2, and a pseudo-bulk model for particle growth to describe the microgel size (distribution) in Chapter 3 were presented. This chapter introduces a hybrid Kinetic Monte Carlo model to enable a depiction of the internal particle structure. This model complements the previously introduced models, thus the key properties of the microgel synthesis - polymerization progress, microgel growth and particle network structure - can be simulated.

4.1 Introduction and literature review

In Monte Carlo (MC) simulations, the sequence of events in a control volume is calculated based on a randomized selection of the events according to their respective probabilies. Hence, MC can be applied to simulate the stochastic nature of reaction and aggregation events (Kremer & Binder (1988), Platkowski & Reichert (1999), Lattuada et al. (2004)).

MC simulations are quite established to track the structure of a polymer network. Therein, the reactions are calculated individually, allowing to track the desired property, e.g., the composition, chain length, cross-links, and branching points for each individual polymer chain. A polymer network then consists of several (cross-linked) polymer chains with the respective chain length and cross-linking distribution (e.g., Tobita & Yanase (2007); Wang et al. (2017)).

The level of detail of this information allows to express the impact of the polymer chain properties on the reaction kinetics. A common example is the dependence of the termination reaction on chain length, when diffusion of larger chains and in consequence intermolecular termination is limited. Further, with the inclusion of branching and cross-linking reactions, the cross-linking density (Hamzehlou et al. (2012, 2013)) and beyond that the gelation of the system can be predicted (Šomvársky & Dušek (1994); Tripathi et al. (2014)). For PNIPAM-based microgels, a MC study was presented to elucidate the effect of reaction kinetics on the lengths between cross-links and length of dangling chains and therefore resulting core-shell structure (Gavrilov et al. (2020)). Other references use the comprehensive description of the polymer from MC simulation to predict inhomogenities in a system (Lemos et al. (2015)) or use the architectural information to derive polymer product properties (Agrawal & Agrawal (2018)).

The high level in detail of MC simulations increases the computational cost. To reduce computational cost of calculating each reaction one at a time, hybrid approaches can be employed. Only less frequent reactions are calculated stochastically, while faster reactions, such as the propagation, are calculated from differential equations (e.g., Tobita & Yanase (2007),Tripathi et al. (2014)). Alternative approaches to decrease the computational cost suggest sequencing of polymer chains for tracking the copolymer composition (Wang & Broadbelt (2011)) or calculation from binary trees (Chaffey-Millar et al. (2007); Nasresfahani & Hutchinson (2018); Van Steenberge et al. (2014)). A comprehensive review on Monte Carlo simulation for polymer systems is provided by (Meimaroglou & Kiparissides (2014)).

When describing disperse polymerization systems, like precipitation polymerization, the ability to describe compartmentalization is another benefit of the MC method. The particles are treated as nano-reactors for the polymerization reactions. In consequence, reactions of radicals are limited to the molecules in the same particle and mass transfer with the continuous phase, while reactions with radicals of other particles are inhibited.

The polymerization progress of disperse systems is dominated by two competing mechanisms (Tobita & Yanase (2007)): On one hand, compartmentalization isolates radicals, which limits termination reactions therefore accelerating the overall polymerization rate. On the other hand, when a second radical enters the particle, the confined particle volume, especially for smaller particles, leads to rapid termination reactions. Polymerization comes to a hold in the particle until another radical enters and in consequence, the overall polymerization reaction descreases. The discrete formulation of the MC simulation does not require limiting assumptions such as the maximum number of radicals in zero-one-models, or averaged radical numbers in particles in pseudo-bulk models. Instead, the number of radicals in a particle depends on the events taking place in the individual particle (e.g., termination) and radical entry and desorption events with the surrounding continuous phase.

For the precipitation polymerization of PNIPAM- and PVCL-based microgels, Maldonado-

Parra (2019) introduced a MC-based model to capture the effect of compartmentalization of radicals. Simulations with and without oligomer absorption demonstrate that absorption of radicals leads to larger particles and more narrow particle size distributions. Still, the obtained particle size distribution are wider than experimentally observed (cf. Chapter 3.3), which indicates, that aggregation of particles also takes part in microgel formation. This conclusion is also supported by the findings of the pseudo-bulk model in Chapter 3.5.1. Further, for a comprehensive description of the microgel synthesis, the model presented by Maldonado-Parra (2019) also misses the copolymerization mechanism for incorporation of cross-linker, which is a necessity to capture the important network properties.

Moreover, MC simulations can capture the effect of the internal polymer structure on the reaction kinetics and the resulting particle size distribution. The internal polymer structure can be tracked in terms of their properties, such as the polymer chain length, their composition and cross-linking points (Hamzehlou et al., 2014a,b). Mass transfer mechanisms such as radical entry and desorption can be precisely linked to the oligomer length or composition (Drache et al., 2018; Hamzehlou et al., 2014b; Marien et al., 2019a). Hence, chain length distribution as well as local inhomogenities in the copolymer composition can be simulated in detail. Stubbs et al. (2008) even proposed to allocate the shorter polymer chains to local positions and simulate their diffusion within the polymer network. This enables the prediction of the internal morphology. As an example, they illustrate the inhomogeneous radical distribution of polymers formed by chain transfer reactions.

With the information on length and cross-links of each polymer chain, chain length dependent termination reactions can be expressed and thus, diffusion limitation to accelerate the polymerization in the particle and hence particle growth. Calculation of a multitude of particles shows, that these effects can change the particle size distribution over the course of conversion. This ultimately shows the effect of kinetics of individual chain growth on the resulting particle size distribution (e.g., Drache et al. (2018); Marien et al. (2019a,b)).

To provide an accurate representation of the resulting particle size distribution and reduce the impact of single particle results, simulations are generally conducted for multiple particles. However, the number of simulated particles is critical for the calculation time and thereby especially depends on the saved particle information. Hence, the in literature provided numbers for simulated particles vary significantly. These range from ten particles (Stubbs et al. (2008)) for high detail, such as simulating the position of radicals in the particle, to more than 500.000 particles (Drache et al. (2018)) with few simplifications for fast computation, e.g., limitation of the maximum number of radicals per particle. For the investigated systems, the particles are considered to be stable with an initial particle size or size distribution. Their number remains constant throughout polymerization, neglecting formation of new particles or aggregation.

Monte Carlo simulations have also proven to be effective for the solution of population balance equations of disperse systems with aggregation (Khalili et al. (2010); Meimaroglou & Kiparissides (2014)). The population is thereby represented by a smaller subsystem, within which the aggregation events are calculated according to their probability. To calculate the aggregation probabilities, it can be combined with theories regarding particle aggregation, such as the DVLO theory (Gervasio & Lu (2019)). This approach does not require the discretization of the solution domain and captures the stochastic behavior of aggregation events well. However, the rounding error of the selected subset results in a reduced accuracy. Especially when dealing with sources or loss of particles, such as formation or aggregation, it appears to be beneficial to adjust the size of the control volume in order to have either a manageable number of particles in regards to computational cost or a sufficient number of particles to provide a good representation of the particle size.

In conclusion, Monte Carlo simulations bring multiple benefits for the simulation of the investigated precipitation polymerization of microgels: The simulation of individual reactions, one at a time, provides a detailed representation of the polymer network. This provides insight into the cross-linking distribution of the formed microgels. The discrete treatment of compounds within the reaction volume enables an efficient description of compartmentalization in heterogeneous systems, such as precipitation polymerization. Furthermore, applying Monte Carlo methods to aggregation events of polymer particles, a representation of a resulting polymer morphology can be simulated. So far, no modeling approach has been presented for application of MC simulation to precipitation polymerization, combining the aforementioned advantages for aggregation and network formation. However, the ability to simulate the discrete formation of new particles by precipitation, the transition of zero-one kinetics from precursor particles to pseudo-bulk kinetics from larger microgels and the ongoing growth with discrete number of radicals over the full range of conversion should be an explicit strength of the MC simulation.

This chapter introduces a Monte Carlo-based modeling approach for the batch synthesis of PVCL-based microgels from precipitation polymerization. The model therein combines three scales of the microgel formation: the mass balance within the reactor on the macro-scale, the aggregation of precipitating and growing microgels on the meso-scale, and the individual reactions within microgels on the micro-scale. This provides a prediction of the overall polymerization progress, such as the process model in Chapter 2, and the particle growth and size distribution, such as the population balance equation model in Chapter 3. Moreover, the Monte Carlo simulation enables a detailed description of the internal microgel structure.

For the two coarser scales, the macro- and the meso-scale, the simulation results can be validated with experimental data. Therein, the parameter values are used as estimated with the previously introduced models. On the macro-scale, the simulation results are compared to experimental data from reaction calorimetry as a measure for the overall polymerization progress. On the meso-scale, the simulated final microgel radius and size distribution are compared to the experimental data. Thereby, the validation shows a deviation between experimental data and simulation due to the incompatibility of the model with the employed parameter values.

However, a better agreement of experimental data and simulation can always be accomplished by adjustment of parameter values. Thus, the presented approach is to be understood as a tool for the prediction of the microgel architecture, once sufficient parameter values are provided. To illustrate the potential of the model, the focus is set on depiction of the construction of a microgel particle form aggregation and the internal cross-link distribution.

4.2 Hybrid kinetic Monte Carlo model

The hybrid kinetic Monte Carlo (KMC) model combines stochastic Monte Carlo method with differential algebraic equations (DAE). While the Monte Carlo simulation performs well for smaller subsystems, the computational cost increases with the size of the simulated control volume. Thus, to capture the entirety of the reaction system beyond the simulation of single particles, the KMC model is paired with in differential algebraic equations for mass balances and precipitation.

The modeling approach distinguishes three different scales in the process of microgel formation. These scales are illustrated in Figure 4.1. The macro-scale describes the reactor scale. This scale comprises the differential equations for the reactor content for initiator and comonomer as well as balances for radical species in the liquid phase. The radical balances determine the number of precipitating radicals, their respective composition and species. This information is passed to the meso-scale. The meso-scale describes the interactions among the particles. It entails a population balance for the number of generated particles within a smaller control volume. Aggregation of the particles, which is simulated by MC simulation, changes their number and respective composition, providing the alteration in particle size distribution. The adjusted subset of particles is passed to the micro-scale. The micro-scale describes reactions within a growing microgel. Every particle is considered to operate as a nano-reactor (Drache et al. (2018)). The polymerization reactions in the respective particles are simulated by the KMC method, which ultimately provides the comonomer conversion per particle. The monomer consumption combined with the respective number of particles is returned to the macro-scale in form of the reaction rates of initiator and comonomers. Hence, the three scale are intertwined by the concentrations of monomers and oligomers as well as the number of precursor particles and microgels.

The stochastic elements in meso- and micro-scale are based on the direct method for the

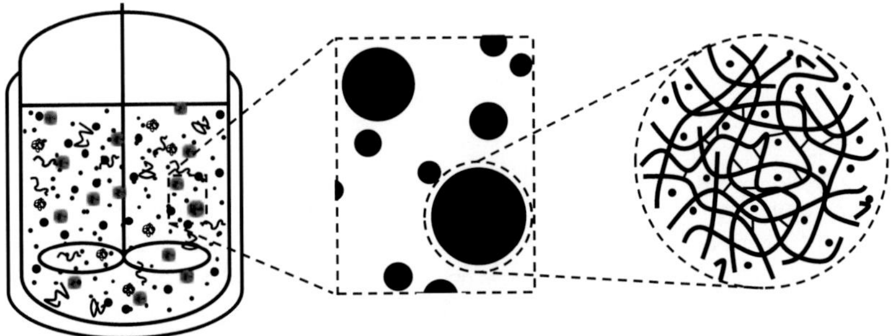

Figure 4.1: Illustration of the three scales of the hybrid KMC model: the macro-scale of mass and energy balance of the entire reactor (left), the meso-scale for particles, their growth and aggregation (center), and the micro-scale for the reactions within particles and the particle structure (right).

KMC simulation developed by Gillespie (1976). A fundamental assumption to the KMC simulation is that only one event occurs at a time. Each of the potential events M can occur with the respective probability a_ξ. The probability depends on the number of component combinations in terms of absolute numbers (molecules or particles, respectively) $\mathcal{H}_\xi$ and the reaction parameter C_ξ. Then, the first random variable r_1 determines the occuring event ξ, and the second random variable, r_2, determines the time τ of the occurring event, based on the sum of all probabilities a_0 of the competing events.

$$a_\xi = \mathcal{H}_\xi C_\xi \ (\xi = 1, 2, ...\mathrm{M}) \tag{4.1}$$

$$a_0 = \sum_{\xi=1}^{\mathrm{M}} a_\xi \tag{4.2}$$

$$\sum_{\zeta=1}^{\xi-1} a_\zeta < r_1 a_0 < \sum_{\zeta=1}^{\xi} a_\zeta \tag{4.3}$$

$$\tau = \frac{1}{a_0} \ln(1/r_2) \tag{4.4}$$

r_1 and r_2 are random numbers from the uniform distribution in the unit interval. The full derivation of the direct method is described in detail in Gillespie (1976) and in literature employing the KMC (Hamzehlou et al. (2014a,b); Lemos et al. (2015); Tripathi et al. (2014)).

In terms of microgel formation and polymerization reaction mechanisms, the hybrid KMC model is based on the same assumptions as for the models described in Chapters 2 and 3. Further, the following assumptions apply to simplify the model:

- Radical entry is limited to radicals with critical chain length. The reaction kinetic for absorption of precursor particles corresponds to the kinetic of radical entry described in Chapter 3.2.2.
- Radical desorption from microgels is neglected.

In the following, the individual model scales are described from coarse to fine, with an explanation of the approach to track individual particles and their structure, followed by a detailed description of connection of the model scales to construct the hybrid KMC algorithm.

4.2.1 Macro-scale

On the macro-scale, the reactor is described by the overall differential mass balances for initiator, monomer, and cross-linker. Initiator, monomer and cross-linker are consumed in both the continuous liquid phase as well as in the growing microgels. Between the continuous and the disperse phase, phase equilibrium for these three components applies. The model equations are introduced in Chapters 2 and 3 and given in Appendix E.

Besides the mass balances for initiator and comonomer, the concentrations of radicals in the liquid phase are calculated by differential equations. With these, the formation of oligomers with the critical chain length η is balanced, which provides the rate of precursor particle formation. Integration of this rate over a finite time step Δt provides the number of precursor particles formed within this time frame. These precursor particles are attributed to the same generation. The precursor particles are distinguished according to their species of terminal radicals, so for the investigated copolymerization, two kinds of particles are formed. Each particle species is characterized by the number of monomer units, which corresponds to η, and the number of PDB available for cross-linking. The number of precursor particles of each species in the generation is then passed to the meso-scale to calculate particle aggregation.

4.2.2 Meso-scale

The stochastic nature of particle coagulation is described by KMC simulation. Therefore, a representative control volume V_c with a discrete number of particles is specified (Gervasio & Lu (2019)). The particles are distinguished in terms of their generation g and species s. Their respective number is stored in the $s \times g$ matrix N_c. Within the control volume, all particles can coagulate, and two aggregating particles form a new particle.

The aggregation probability of two particles is based on the formulation in Chapter 3.2.3 which is based on the formulations by Araújo et al. (2001) and Vale & McKenna (2009). It is calculated by

$$a_{s,g,\tilde{s},\tilde{g}} = \begin{cases} \beta(r,\tilde{r}) N_{csg} \left(N_{c\tilde{s}\tilde{g}} - 1\right)/2, & s = \tilde{s} \wedge g = \tilde{g} \\ \beta(r,\tilde{r}) N_{csg} N_{c\tilde{s}\tilde{g}}, & \text{else.} \end{cases} \tag{4.5}$$

with the coagulation kernel β. The formulation for the coagulation kernel is equivalent to the formulation in Eq. (3.20), where coagulation of two particles with both radii above the stable particle radius r_{stable} is prohibited. One exception of the aggregation probability applies for the precursor particles of the latest generation gc with any other particle. To represent their rapid aggregation and since radical entry of oligomers is neglected, the coagulation kernel is calculated from the radical entry mechanism (cf. Chapter 3.2.2)

$$\beta(r,\tilde{r}) = \begin{cases} 0, & r \geq r_{\text{stable}} \wedge \tilde{r} \geq r_{\text{stable}} \\ \frac{f_e 4\pi r D_{W_i}}{\sqrt{\eta}}, & \tilde{g} = gc \\ \frac{f_e 4\pi \tilde{r} D_{W_i}}{\sqrt{\eta}}, & g = gc \\ \frac{2k_B T N_A}{3\mu^l W}\left(2 + \frac{r}{\tilde{r}} + \frac{\tilde{r}}{r}\right), & \text{else.} \end{cases} \tag{4.6}$$

For the randomly selected aggregation event, its duration is determined according to Eq. (4.4).

When two particles coagulate, the compositions of the aggregating particles are combined in a new particle. The number of particle species in N_c for both previous particles are each reduced by one. The newly formed particle is allocated in the generation of the older particle, as the older microgel "absorbs" the younger microgel. When the same particles aggregate repeatedly, the number of the newly formed particle increases correspondingly.

The coagulation continues until the time step Δt is completed. The new number of particles and their respective composition of the particle species is then passed on to the micro-scale for a simulation of the polymerization within the particles for the same time step Δt.

4.2.3 Micro-scale

The micro-scale comprises the reactions within each particle by KMC simulation. This requires the composition of the reacting compounds in each control volume, which correspond to the respective particle volumes (Tobita & Yanase (2007)). The polymer composition includes the absolute molecule number of initiator I, initiator radicals $I^\bullet$, monomer and cross-linker M_i

inside the microgel, the number of active radicals λ_i of the species i, the number of dissolved radicals $R_{1,i}$ formed by chain transfer to monomer, as well as the numbers of available PDB and formed cross-links X. In addition, $M_{\text{reac}i}$ counts the number of monomers bound into the polymer in the time step Δt to calculate the monomer consumption in the particle.

According to the KMC method in Eqs. (4.1) - (4.3), the reaction channel ξ occurs in a particle. Then, the composition of the particle must change accordingly. These changes in composition are listed for the respective reactions ξ in the columns of the reaction channel matrix ν in Table 4.1. Corresponding to the occurred reactions, the required time $\tau_{\text{KMC}gs}$ for the event in a particle of generation g and species s is calculated by Eq. (4.4). The simulation time for the particle $t^{\text{p}}_{\text{KMC}gs}$ is updated by $\tau_{\text{KMC}gs}$ correspondingly. The simulation time is treated for each particle individually and the KMC simulation is repeated until the time step Δt is completed for every particle.

When the simulation step is completed for all particles, the reaction rates for monomer consumption $\mathcal{R}_{\text{M}_i}$ in all particles during Δt is calculated with the number of the respective particles $N_{\text{c},gs}$ of generation g and species s by

$$\mathcal{R}_{\text{M}_i} = \sum_{g,s} \frac{N_{\text{c},gs} \text{M}_{\text{reac}gsi}}{t^{\text{p}}_{\text{KMC}gs}}. \tag{4.7}$$

The reaction rate is then passed back to the macro-scale, to calculate the overall change of the monomer concentration.

4.2.4 Polymer structure

To document the growth of the polymer structure, some information is collected during simulation. The focus is set on the particle structure as a result of particle aggregation and cross-link distribution. The information is selected for a limited number of particles, determined by their initial generation, to reduce the computational cost. These particles are tracked from their birth throughout the polymerization and aggregation until the end of the simulation.

For this limited particle set, the following information is saved:

- *Radicals:* Indices of the respective species of the primary polymer chain radical in the microgel (row vector),
- *Length:* Length of the respective primary polymer chains (row vector),
- *Position:* Position of the radical in terms of the radius with respect to their original particle's center (row vector),
- *FreePDB:* Available PDB in the respective polymer chains (row vector),

Table 4.1: Reaction channel matrix ν for KMC algorithm. Indices i, j for the reactant combinations represent the species of (co-)monomer or cross-linker (VCL, NIPAM, BIS). The entries of ν correspond to the reaction mechanisms in Table 2.1 and Table 3.1.

Species	Reaction ξ						
	initiator decomposition d	initiation I_i	chain propagation p_{ji}		chain transfer to monomer fm_{ji}	chain termination td_{ji}	cross-linking x_{ij}
				a			
I	-1	0	0	0	0	0	0
$I^{\bullet}$	2	-1	0	0	0	0	0
λ_i	0	1	1	1	0	-1	$-\delta_{BIS,j}$
λ_j	0	0	-1	0	-1	-1	$\delta_{BIS,j}$
P	0	0	0	0	1	2	0
PDB	0	$\delta_{BIS,i}$	$\delta_{BIS,i}$	$\delta_{BIS,i}$	$\delta_{BIS,i}$	0	$-\delta_{BIS,j}$
X	0	0	0	0	0	0	$\delta_{BIS,j}$
R_i	0	0	0	0	1	0	0
R_j	0	0	0	-1	0	0	0
M_{reaci}	0	1	1	1	1	0	0
X_i	0	0	0	0	0	0	1

[a] with oligomer of length 1

- *Oligomers:* Oligomer species and the row index for the chain it was formed by to retrieve the chains position (matrix with 2 columns),
- *Agglomerate:* Agglomeration information including the Cartesian coordinates of center of the absorbed particle where the chain is located in (1.-3. column), the radius of the absorbing and the absorbed particle (4.+5. column), the generations of the absorbing and the absorbed particle (6.+7. column) and the indicator whether the chain was formed by initiation (0) or included by absorption of a particle (1 or 2, when the particle was tracked previously) (8. column),
- *Polymers:* Structure containing composition information for each chain with:
  - *Cross-linker:* Cross-linker units in the chain with their position (1. column) and whether the PDB is cross-linked (1) or not (0) (2. column),
  - *Links:* Cross-link units, with the unit number of the cross-link (1. column), the number of polymer chain of the PDB (2. column), the position of the reacted PDB (3. column),

When the respective reactions on the micro-scale occur in the particle, the polymer structure changes accordingly.

Another aspect is the particle structure due to aggregation. When two particles agglomerate, the information of the respective particles is merged and tracking is transferred to the newly formed particle. Since only a small number of particles is tracked, the particle structure of one of the aggregating particles is not saved. In this case, the structure information of the "unknown" particle is constructed based on the available particle properties from the micro-scale. These particles are constructed with even radical length and PDB distribution. Information, which cannot be approximated from the general particle information, such as cross-linking distribution, remains unknown and is set to zero.

In principle, tracking of the particles follows the absorbing particle, hence the "older" particle. Hence, tracking is simply transferred to the newly formed particle. In the case the absorbed particle was tracked, the coordinates of its center change, so the coordinate system origin remains the center of the currently tracked particle. For the aggregation, the radius in the instance of aggregation of the absorbing and the absorbed particles as well as their respective generations are saved.

4.2.5 Hybrid model algorithm

The three scales and the calculation of the polymer structure are simulated sequentially. Similar approaches were proposed previously: Stubbs et al. (2008) presented an algorithm to

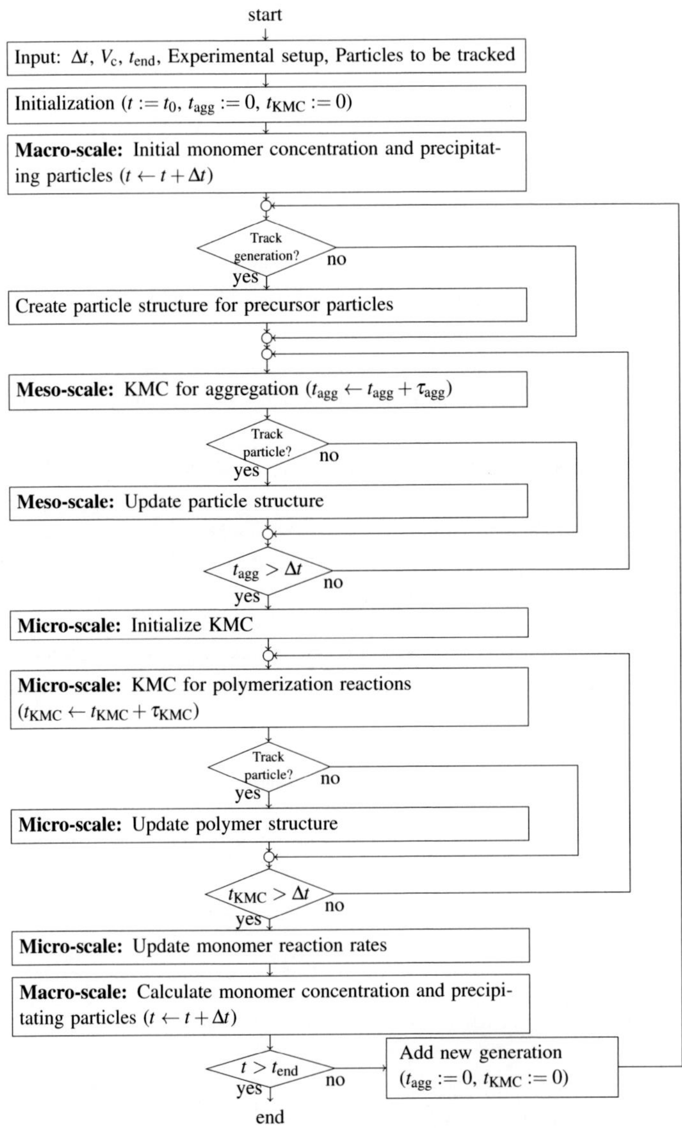

Figure 4.2: Scheme for the algorithm of the hybrid KMC model.

combine the liquid ODE based formulation with the particle internal KMC simulation. Yet, the presented algorithm does not account for precipitation and aggregation, as a constant number of particles is simulated. Likewise, the algorithm by Maldonado-Parra (2019) combines a ODE based formulation for the liquid phase with KMC simulation for the particle internal reactions. While this model also accounts for particle formation, particle aggregation is neglected. In the following, an adjusted algorithm for the simulation of the hybrid KMC model with the three previously described scales is used. The algorithm is illustrated in Figure 4.2.

When the algorithm is started, it receives the input information on the step size Δt, the control volume V_c, the simulation time t_{end}, and the experimental setup in terms of concentrations of the reactants. In addition, it receives the information which particles should be tracked in terms of their generation and the number of particles per generation. Then, the simulation is initialized by setting the simulation times t, t_{agg} and t_{KMC} to zero and providing the initial conditions for the reactant concentrations. These are the initial values for the mass balances on the macro-scale for the initial time step Δt. The calculation of the macro-scale provides the first generation of precursor particles and the updated initiator, monomer, and cross-linker concentrations. When particles of the first generation should be tracked, the particle structure of the precursor particles is created. Otherwise, this step is skipped.

Number and species of precursor particles are passed to the meso-scale to calculate the aggregation. For the aggregation event, the aggregation step τ_{agg} is added to the internal simulation time t_{agg}. After each particle aggregation, the particle structure is updated for the tracked particles. These two steps are repeated until t_{agg} exceeds Δt. Then, the updated set of aggregated particles is passed to the micro-scale.

On the micro-scale, initially, the numbers of initiator, monomer, and cross-linker molecules in the particles are calculated from the latest concentrations from the macro-scale and the particle volumes. Then, the reactions in the particles are calculated with the KMC. For simplicity, this is only expressed for an individual particle here. However, calculations are performed for all particles. Each particle obtains an internal simulation time t_{KMC}, which is updated according to the calculated reaction duration τ_{KMC}. For the tracked particles, the polymer structure is changed accordingly. As long as t_{KMC} is below Δt, the KMC on the micro-scale is repeated. A particle is excluded from further reactions, when its corresponding t_{KMC} exceeds the time step Δt. When all particles completed the simulation time Δt, the reaction rates for initiator, monomer, and cross-linker consumption in the particles are calculated by Eq. (4.7).

The reaction rates are passed to the macro-scale and the differential equations for the mass balances are solved for the time step t to $t + \Delta t$. The precipitated particles from this time step are attributed to the next generation. While the simulation time t is below the intended simulation time t_{end}, the algorithm proceeds with aggregation for the next time step Δt. When the simulation time t reaches t_{end}, the simulation terminates and the results are saved.

The number of iterations of the outer loop, which passes macro-, meso-, and micro-scale, is known a priori as it is inversely proportional to the step size Δt. For the inner loops in meso- and macro-scale, the number of iterations depend on the particle number, and thus, V_c, and the reactive species, respectively. Thereby, τ_{agg} and τ_{KMC} increase with decreasing number of particles and reactive species within the particles. Towards the end of the polymerization time, when particle number and reactive species within the particles decline, this leads to an acceleration of the algorithm and convergence of the algorithm is guaranteed.

The model is implemented in Matlab 2019a. The files are available for download [1]. Parameter values for reaction rate constants for propagation and chain transfer to polymer have been used from quantum mechanical calculations listed in Table A.2 and Table A.3. The reaction rate constants for termination reactions have been used as estimated with the terpolymerization model (cf. Table A.6). General parameter values and constants are transferred from the previous models and used as listed in Table A.4. The time step Δt is 0.25 s and the control volume $V_c = 5 \cdot 10^{-18} m^3$.

4.3 Results and discussion

This chapter summarizes the simulation results for the synthesis of cross-linked PVCL-based microgels of the hybrid KMC model. For better comparability to the results presented in Chapters 2 and 3, the simulated experimental setup corresponds to the reference experiment with 2.5 mol-% BIS. The simulation results are first compared to the experimental data used in Chapter 3 for validation. Then, the predicted particle structure in terms particle aggregation and cross-linking distribution are discussed.

4.3.1 Prediction of reaction enthalpy transfer rate, average microgel radius and PSD

The simulation results can be validated on the macro- and the meso-scale. Validation is possible with the overall reaction progress, in terms of the reaction enthalpy transfer rate, and the microgel growth, in terms of the average microgel size (distribution). Figure 4.3 shows the predicted reaction enthalpy transfer rate from the hybrid KMC simulation over polymerization time. The experimental data of the reference experiment (cf. Figure 3.2) is shown for comparison. The simulation results show the characteristic profile with a first maximum at approx. 25 s and a second maximum after approx. 100 s. In Chapter 2, it has

[1] https://doi.org/10.18154/RWTH-2020-09709

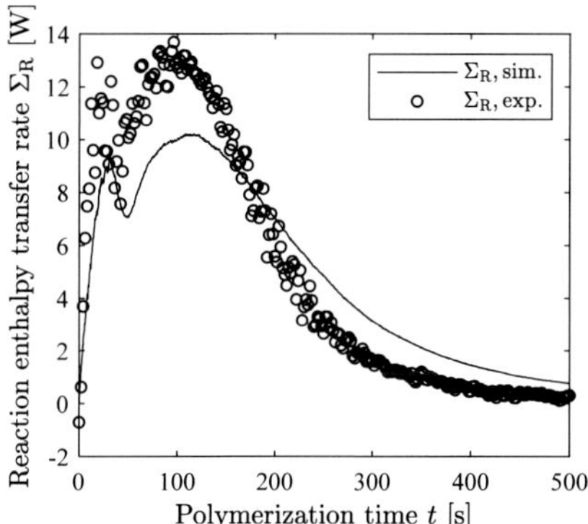

Figure 4.3: Reaction enthalpy transfer rate calculated with hybrid model compared to experimental data (cf. Figure 3.2). For improved readability, error bars are not shown.

been explained that this characteristic is related to the inhomogeneous consumption of VCL and BIS. Therefore, the model also successfully depicts the inhomogeneous consumption of VCL and BIS.

However, there is a clear deviation between the simulated and the experimental profile. The slope of the simulated reaction enthalpy transfer rate Σ_R is significantly lower and the asymptotic value is higher. This lower slope is caused by a slower consumption of monomers in simulation, as Σ_R depends on the reaction rates. Hence, the overall polymerization progress calculated with the hybrid KMC model is too slow, which results in the observed deviation of the slope.

This deviation can directly be related to the parameter values: the parameter values are neither estimated nor adjusted with the hybrid KMC model, but, as pointed out earlier, transferred from the previous models described in Chapters 2 and 3. While the parameter values are the same, the model implementation differs: the hybrid KMC model describes the zero-one kinetic of the earlier stages of the polymerization, while the pseudo-bulk model assumes an average radical number for particles of the same size. This simplification is known to lead to inaccuracy for smaller particle sizes, which affects the parameter estimates, eventually causing the discrepancy. An adjustment of the fitted parameter values could reduce the deviation among experimental and simulated data.

The simulation results provide the respective sizes of the microgels in the control volume. Figure 4.4 shows the average radius for microgels in V_c over polymerization time. Within

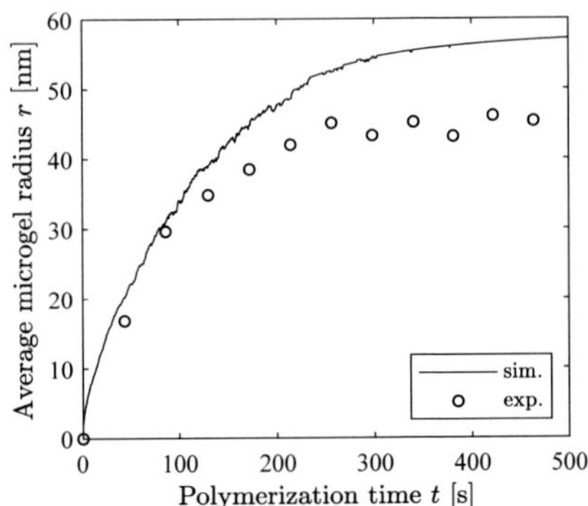

Figure 4.4: Average microgel radius over polymerization time, calculated with hybrid model ("sim."), in comparison to *in situ* DLS measurements of the microgel radius ("exp.").

the first 200 s after initiation, the mean radius grows rapidly. Then, the average growth of the microgels slows down and at the final simulation time of 500 s, the mean microgel radius is 54 nm. For comparison, the experimental data from *in situ* DLS measurements are provided (cf. Appendix D.1). A deviation between experimental and simulation data can be observed. However, any conclusion based on this deviation with regard to the model quality should be done carefully. The *in situ* DLS measurements are known to have a large margin error (cf. Appendix D.1), because the extraction process affects the microgel size. The measurements predict smaller microgel radii compared to the redundant DLS measurements, and thus, the measurements can only serve for a qualitative comparison. The time evolution predicted from the model qualitatively matches the experimental measurements: A first rapid growth phase within the first first 200 s, followed by a slow growth phase where the particle radius increases only slightly until the end of simulation time. Overall, the predicted mean particle radius at 500 s is slightly smaller than final microgel determined experimentally by DLS (cf. Figure 3.3, approx. 60 nm).

The experimental setup was already simulated using the pseudo-bulk model in Figure 3.3. Compared to the growth in Figure 4.4 by the hybrid KMC model, the prediction within the first 200 s with the latter is less steep. The cause for this difference can be best explained by the particle size distribution over time, predicted by the hybrid KMC model. Figure 4.5 shows the PSD for different times in terms of microgel numbers in the control volume over their respective radius. Each time shows an individual and distinct peak, thereby increasing reaction times lead to peaks at larger average microgel radii. For early polymerization times,

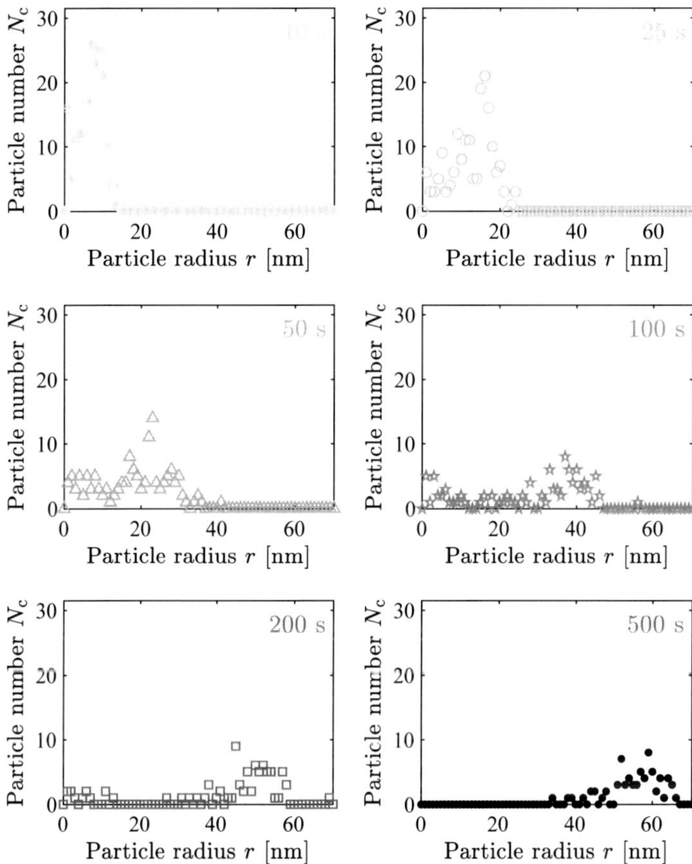

Figure 4.5: Particle size distribution calculated with hybrid model for different polymerization times (10 s, 25 s, 50 s, 100 s, 200 s, and 500 s). The particle number is calculated by partitioning the radius in bins of 1 nm and categorizing the particles according to their radius.

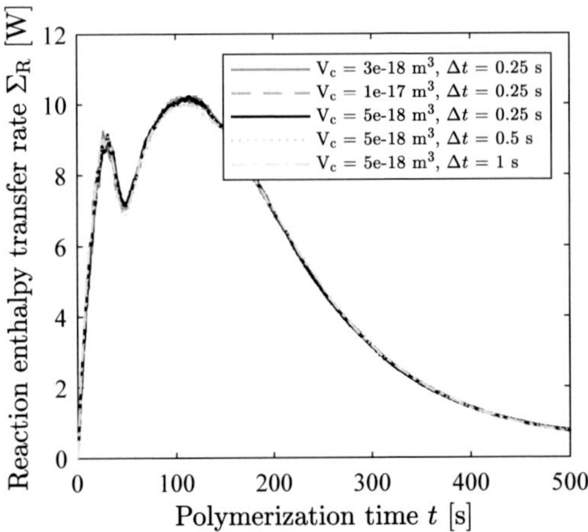

Figure 4.6: Comparison of simulation runs with different control volumes and step sizes in terms of the reaction enthalpy transfer rate.

such as 10 s after initiation, a high number of small particles is present. For the selected control volume V_c, the maximum number of particles in the volume is 212 (particles *after* aggregation calculation, cf. Figure 4.2). With ongoing polymerization time, the number of particles declines, while the mean of the particle size distribution shifts to larger radii. From 150 s, the size distribution for larger microgels is almost constant, shifting only with the mean radius to higher radii. However, for the size distribution for t =150 s, there are also particles present with smaller radii (< 10 nm), which lower the average microgel radius.

This is a fundamental difference to the results in Figure 3.4. The smaller particles are caused by the different implementation for radical entry and aggregation of precursor particles in the hybrid KMC approach. Here, precursor particles continuously precipitate, while in the pseudo-bulk model, precipitation is inhibited after a short nucleation phase of approx. 10 s, as radicals are absorbed by growing microgels before they obtain the critical chain length for precipitation. Nonetheless, both approaches come to a similar conclusion, as the precipitated precursor particles in the hybrid KMC model also predict an absorption of the smaller particles. With ongoing polymerization time, this leads to a monodisperse PSD with 72 particles in V_c with a single peak around 58 nm.

It should also be highlighted that the results for Σ_R and the radius show only little sensitivity towards the time step Δt, which connects the different reaction scales, and the control volume

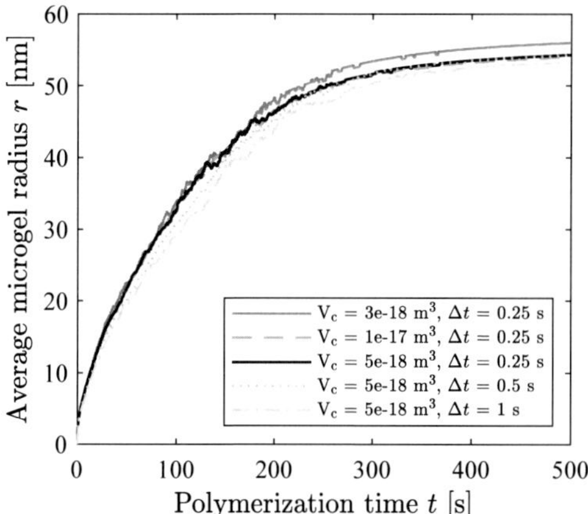

Figure 4.7: Comparison of simulation runs with different control volumes and step sizes in terms of the average radius.

V_c. Comparisons of the simulations performed with different combinations of Δt and V_c are given in Figure 4.6 and Figure 4.7. The simulation results for the reaction enthalpy transfer rate show no significant deviations in dependence on V_c or Δt. For the mean radius, slight variations for simulations performed with different V_c are shown. This can be related to the number of particles which are calculated in the control volume V_c, which decrease proportional to the decrease of V_c.
A low number of particles in V_c can lead to two significant drawbacks: First, growth of individual particles has a higher impact on the mean value, and second, the time predicted for an aggregation event to take place exceeds Δt, which suppresses the execution of the aggregation event. Both effects lead to a poor representation of the simulated system. In comparison, the number of reactions in the particles itself for the calculation of Σ_R is much higher, which automatically provides a more accurate prediction of the mean value. With an increase of V_c from 5e-18 m^3 to 1e-17 m^3, the final particle radii mostly align. This shows that with an increase of the control volume, the simulation is in general independent of the choice of the selected parameters Δt and V_c.

For the sake of completeness, it should also be noted that the hybrid KMC model is in principle capable of matching the experimental data with better accuracy when parameter values are adjusted. In Appendix E.2, the results of a simulation performed with manually adjusted parameter values for the diffusion limitation of the termination rate constants in the gel phase ($k^{g}_{t,diff11}$, $k^{g}_{t,diff22}$) are provided. The hybrid KMC model is not suited to be used for a parameter estimation due to its long CPU time. For the simulation of $t_{end} = 500\,s$ performed with a control volume of V_c = 1e-17 m^3 and $\Delta t = 0.25\,s$, the CPU time is 6.46 h, without tracking of any particle structures. Moreover, for a derivative-based parameter estimation, the stochastic part of the hybrid KMC does not allow the calculation of derivatives. Either, an approximation of the derivatives is required or the parameter estimation must be formulated as a derivative-free optimization problem. Alternatively, further adjustment the pseudo-bulk model to match the hybrid KMC model (e.g. in terms of zero-one kinetics) could be performed and the newly formulated pseudo-bulk model is used for parameter estimation.

4.3.2 Microgel structure

In addition to the simulation of monomer conversion and the microgel size growth, the model's purpose is a prediction of the internal structure. In the following, the results for the particle structure for aggregation and cross-link distribution will be discussed.

In the simulation for the reference experiment, particles formed in generations 1, 100, 400 and 800 (precipitated after 0.25 s, 25 s, 100 s, and 200 s, respectively) were tracked over the progress of polymerization time. The microgel structures for the finally obtained particles as the result of aggregation are shown in Figure 4.8. The outer diameter (black line) represents the final microgel radius. Each gray circle represents an absorbed microgel, precursor particle or absorbed radical. The size of each gray circle corresponds to the radius of the particle, and its radial position corresponds to the radius of the absorbing particle in the moment of aggregation. This represents how the absorbed particles are build in the microgel structure. For the absorbed particles, dark gray represents older generations and lighter gray represents particles of younger generations. Particles, that were tracked but then absorbed by a particle of an older generation - and hence tracking was transferred to the absorbing particle - are identified by a dashed line.

The particle in the upper left corner was formed in the first generation and tracked over the full polymerization time. Thus, since this particle is tracked over the longest lifespan, it provides the most information on the absorption and aggregation. Primarily, smaller particles are absorbed. These are absorbed continuously, as they are distributed evenly across the radius. The visualization shows that along its growth up to a radius of 53 nm, only few larger particles were absorbed, as the larger, light gray circles illustrate. Especially when approaching the outer radius during growth, absorption of larger particles declines.

The remaining particles were formed in generations 100, 400, and 800. Ultimately, the particles are absorbed by different particles and are located in particles of generations 2, 49,

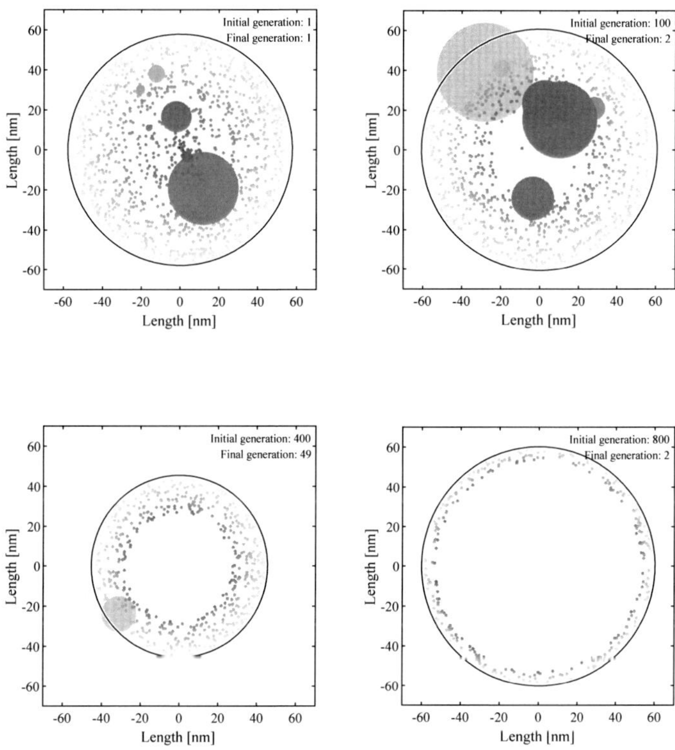

Figure 4.8: Microgels constructed of aggregated particles. Each circle represents an absorbed microgel. Dark gray circles thereby present absorbed microgels of earlier generations. Circles with dotted lines represent microgels that were tracked previously and then absorbed by a growing microgel of an earlier generation. The tracked particles precipitated in generations 1, 100, 400 and 800 (*initial* generation) and were transferred by absorption to other particles, ultimately ending in particles of generations 1, 2, 49, and 2, respectively.

and 2, respectively. As the absorbing particles are not tracked, the centers of the absorbing particles remain blank. Only particles which were absorbed after the tracked particle are depicted. At these late stages of the microgel growth, no larger particles are absorbed anymore.

These results agree with experimental observations of spherical particles: The growing microgels aggregate only in earlier stages of polymerization. An aggregation of larger particles in an advanced stage of polymerization would result in an uneven shape due to the stiffness of the cross-linked polymer networks. The primary absorption of precursor particles or radicals introduces individual radicals which continue to grow on the outer layers of the absorbing microgel, leading to an even, spherical growth.

For the same particles, the cross-link distribution was tracked. The resulting depiction of the cross-link distribution is shown in Figure 4.9. Therein, every black point represents one cross-link that was formed while the particle was tracked. The respective position is composed from the location of the involved PDB in its chain, the location of the chain in its subparticle, and the position of the subparticle. This provides a general though fragmented insight on the cross-link distribution.

For the two particles of initial generation 1 and 100, a radial inhomogeneous cross-linking profile results. The cross-links are located primarily in the center of the microgels, within a radius of approx. 25 nm. In the outer spheres, no cross-links are located. For the particles of initial generation 400 and 800, no cross-links are saved and hence, depicted. This represents that no further cross-links were formed after the particles were absorbed by the larger particle. Hence, the depiction gives insight, that cross-links are formed not only primarily in the center of the microgels, but cross-linking is completed while the microgels still continue to grow by polymerization of the monomer VCL. This results in the previously described heterogeneous cross-linking structure of microgels with a densely cross-linked core and an outer sphere of dangling polymer chains of pure VCL.

In Chapter 2, an initial rough prediction for the internal particle structure from the reaction rates and the predicted growth of the gel volume phase was provided (cf. Figure 2.10, 2.5 mol-% BIS). The result of this coarse and purely qualitative prediction of the cross-linking density comes to a similar conclusion as the detailed simulation with the hybrid KMC model. Both predict a high cross-link density in the central towards the center, and a strongly reduced cross-link density towards outside.

Thus, the results of the hybrid KMC model validate the qualitative insight into the particle structure by the process model. Moreover, the hybrid KMC model predicts the cross-linking density as a result of the cross-linking event in combination with the particle size and the location of the polymer chain within the particle for a more accurate insight into the particle structure.

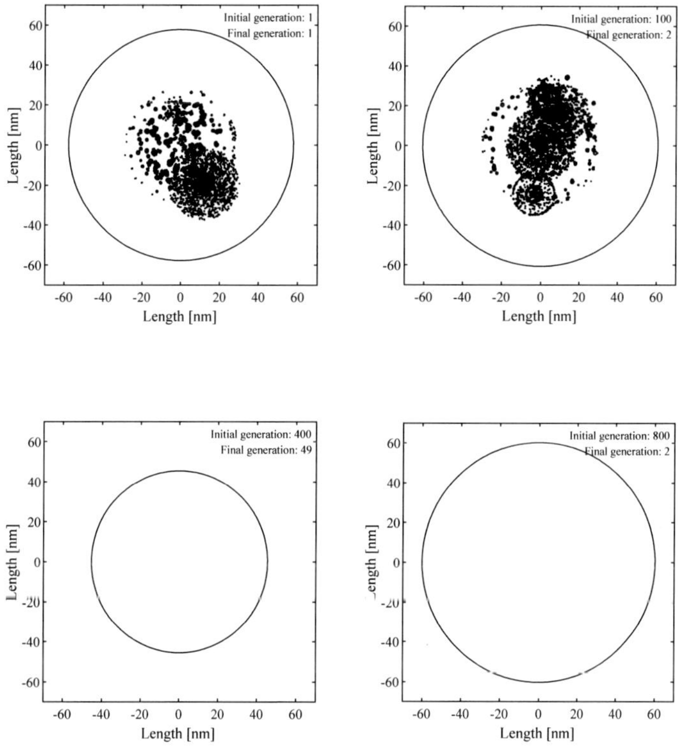

Figure 4.9: Visualization of the final cross-link distribution for the tracked particles. The same particles are depicted as in Figure 4.8.

In conclusion, the hybrid KMC provides the possibility to depict the internal particle structure in great detail. All aggregation and reaction events for all particles in the control volume are calculated explicitly. In the presented results, only a very limited number of four particles is tracked with their particle structure. This is due to the rapid increase in CPU time, when particle structures are tracked in addition to the simulation of reaction progress. The tracking of only few particles means that in terms of their predicted structure, voids remain. E.g., for the formation of cross-links, only those are recorded, which are formed while the particle is tracked. When a non-tracked particle is absorbed or tracking is transferred to another particle, the prediction of the particle structure remains unknown for these incorporated particles within the final particle. To fill these voids and obtain complete visualizations of the final particle structures, it is required to capture the particle structures for all particles in the control volume at any simulation time. This would require the development of a more advanced and efficient storage of the particle properties. The complete calculation of all aggregation and reaction events of the presented hybrid KMC model in this chapter provides the necessary framework.

4.4 Conclusion

A hybrid KMC model for the precipitation polymerization of PVCL-based microgels was introduced. The model describes the three scales of microgel formation: the macro-scale, which comprises the reactor with the continuous liquid phase, the meso-scale with aggregation of growing microgels, and the micro-scale, which describes the polymerization reactions within the growing microgel. While the macro-scale is represented by differential algebraic equations, the meso- and micro-scales are represented by a kinetic Monte Carlo simulation to cover the stochastic nature of aggregation and reaction events. For the stochastic simulations, a representative control volume is selected. The particles within the control volume are simulated with all aggregation events among another and the reactions within. The three scales are intertwined by the number of microgels and the reaction rates of (co-)monomer consumption and a tailored algorithm manages a coordinate solution of the system.

The reaction progress predicted by the hybrid KMC model is slow compared to experimental data from reaction calorimetry and DLS. However, qualitatively a good representation of the inhomogeneous consumption of the monomer VCL and the cross-linker BIS is accomplished. Therefore, the deviation could be directly related to the parameter values transferred from the previous models. Better agreement can be achieved by adjusting the parameter values to match the experimental data. Yet, the presented results illustrate the transferability with reduced accuracy of the model parameters in principle. Thus, the different models can benefit from each other.

With the explicit calculation of aggregation and reaction events, the hybrid KMC model can be employed to provide insight into the particle structure. Two important particle attributes can be investigated: first, the structure of microgels as a result of aggregation and second, the internal polymer network represented by the cross-link distribution.

The particles grow by absorption of radicals and aggregation with other growing microgels. First, the particle structure from absorption of radicals and aggregation of smaller and other growing microgels is investigated. It shows that primarily smaller particles or radicals contribute to the growth of the final microgels. The absorption of larger microgels contributes only to a minor part to the final microgel. This also indicates, that larger microgels grow homogeneously after an initial nucleation phase by absorbing subsequently formed smaller particles, which results in the finally obtained homogeneous PSD. The absorption of smaller particles or radicals instead of cross-linked and hence more rigid larger microgels explains the final spherical shape of the obtained microgels.

The cross-link density within the microgels is also visualized. The prediction shows the inhomogeneous cross-linking distribution within the microgels with an accumulation of cross-links in the microgel center. In the outer sphere, no cross-links are located, which indicates dangling chains consisting of monomer VCL. These comprehensive results agree with the initial rough predictions of the cross-link distribution made with the process model.

For particles that were tracked from an early generation, a quite comprehensive depiction is obtained. Due to the computational cost of tracking the particle structure, only very few particles are calculated. Thus, when a tracked particle aggregates with a larger particle, the internal structure of the second particle is most likely unknown, because this particle might not be tracked. Only an incomplete picture of the aggregate structure can be provided. Hence, an improvement of the efficiency of storage of the particle structure is required to enable tracking of all particles in reasonable computation time and provide a comprehensive prediction of the internal structure.

It should be noted that such an improvement could provide another advantage: Tracking of the internal structure has the potential to link polymer properties and reaction kinetics, such as chain-length dependent termination or gelation. In the indroduced model, simulation of the reaction and tracking of the internal structure are treated separately. Though in principle all particles could be tracked, the number of tracked particles is limited due to computational cost. A computationally more efficient approach would allow to calculate all particle structures within the control volume. This would be the foundation to describe chain-length dependent reaction kinetics or gelation and thus the effect of polymer properties on the reaction kinetics.

With regard to the implemented algorithm, there is potential for improvement in accuracy of the linkage of meso- and micro-scale. In the current implementation, the aggregation events are concentrated in an instant, before the updated particle number is passed to the micro-scale. When the only two radicals in a particle terminate after radical entry, reactions come to a hold until another radical enters, which will happen at the earliest when the time step Δt is completed. Hence, the current formulation should not be completely independent of the choice of Δt. Intertwining both scales, such as passing the concrete moment of aggregation to the micro-scale and then combining the particles on the micro-scale, could resolve this dependence on Δt. The comparison of simulations with different Δt has shown, that this inaccuracy only leads to a marginal error, which would not justify the major revision of the algorithm. However, in case of a comprehensive revision of the algorithm such an improvement should directly be addressed.

In conclusion, the presented model has great potential regarding the insight of the microgel structure. With the explicit calculation of all reaction and aggregation events within the control volume, any particle property can be calculated. Examples are the length in between cross-link units or the fragment length of different (co-)monomers. The model approach can be extended to describe reaction systems with more monomers, such as terpolymerization, to elucidate the structure copolymer-composition. Likewise, the effect of different process operation, such as fed-batch operation, on the particle structure can be investigated, further enhancing the insight of the microgel structure.

Chapter 5

Conclusion and future work

This thesis presents three different models for the precipitation polymerization of cross-linked PVCL-based microgels. The three models focus on different key aspects of microgel synthesis: Progress of the polymerization reaction, growth of microgel, and internal particle structure. While each model provides a different level of detail, information from simpler models are included in the more complex models.

The first model addresses the polymerization reaction kinetics. A two-phase model is used to describe the continuous solvent phase and the disperse microgels. Despite the simplified model, already the large number of unknown parameter values for the fairly unknown reaction system presents a major challenge. While the available experimental data is insufficient for an estimation of all unknown parameter values, it is shown that parameter values from quantum mechanical calculations can successfully be incorporated into the process model to reduce the number of unknowns. The presented process model captures the experimentally determined polymerization progress well, for homopolymerization of monomer VCL and cross-linker BIS, as well as cross-linking copolymerization of VCL and BIS. Simulation results show that the gel phase, representing the polymer particles, serves as the primary reaction locus. The reaction rate constants from quantum mechanical calculations explain the inhomogeneous polymerization progress, showing the faster consumption of cross-linker. The concentration of the cross-linker is too low to be measured in experiments directly, which illustrates the importance of the process model that provides a rough prediction of the inhomogeneous cross-link consumption in PVCL-based and even PVCL-PNIPAM-based microgels. As this model covers the reaction rates and thus is fully capable to describe the copolymer composition, it can be an important tool for the design of experiments for microgels with tailored copolymer composition.

The findings of the first model are transferred into a pseudo-bulk model. This second model

investigates the growth mechanisms during microgel synthesis, providing the microgel size and size distribution. Experimental data, which is used for parameter estimation, is met with high accuracy with the simulation. The estimated parameter values allow direct conclusions on the impact of growth mechanisms. Nucleation only occurs within a short initial time frame. Afterwards, the formed microgels grow simultaneously, as oligomers are absorbed before precipitation. Hence, a uniform particle size distribution is obtained. Though the fitted model meets experimental data, comparison for different experimental setups shows that predictions of the microgel size are difficult. Here, further model development is required to improve the prediction of influences of temperature, initiator and cross-linker concentration.
However, in precipitation polymerization, typically narrow particle size distributions are obtained. Thus, a description of the polydispersity might not always be required. Instead, a reduced modeling approach describing only the average microgel radius, such as proposed by Jung et al. (2019b) might be sufficient. The simpler, less detailed and therefore faster model is better suited for the purpose of process optimization and design of experiments. Nonetheless, when changes of synthesis operation significantly affect the size distribution, a mechanistic explanation as proposed in this model is to be preferred.

The third and most detailed model provides in addition to the polymerization progress and the microgel size distribution the calculation of the internal particle structure. An algorithm is developed which incorporates precipitation in the continuous phase and stochastic simulation for aggregation and polymerization reactions. The stochastic simulation is performed for a representative subsystem of the precipitation polymerization, including each individual aggregation and reaction event, and for a limited particle number, the resulting particle structures are generated. These visualizations reveal that the aggregation of larger microgels has only a minor contribution to the growth, because growing microgels primarily absorb small precursor particles and oligomers. Also, visualizations of the cross-links in the polymer network reveal the inhomogeneous cross-link distribution in the particle. Out of the three, only the third model actually predicts both the microgel size and the cross-link distribution, and hence a comprehensive description of the properties of the formed microgels. This model also gives an accurate description of the compartmentalization of the disperse microgel particles, where the pseudo-bulk is inaccurate for small particles. Thus, full transferability of the parameter values among the models is not given, but better agreement of experimental data and simulation can be obtained when parameter values are adjusted. Due to the long computation time of the stochastic simulation, this model is not suited for applications which require multiple simulation runs, such as parameter estimation, process optimization or design of experiments. In fact, the proposed model provides a framework for the calculation of any polymer properties, including the fragment length of copolymer composition, and chain length distribution. Particle properties, which are difficult to determine by experiment, can be predicted in detail by the proposed model.

The three models are all based on identical assumptions for the reaction mechanisms. This allows to transfer findings from the simplified models to the detailed models. For example, use of parameter values from quantum mechanical calculations or microgel particles as the primary reaction locus are established with the simplest models. Moreover, it enables the transfer of estimated parameter values among the models to avoid the extensive effort of parameter estimation with the complex models.

It should be pointed out that the transfer of estimated parameter values can cause inaccuracies, because minor modifications of the model for each level are inevitable part of the modeling process. For example, the polymer transfer from the liquid and the gel phase in the first and the third model is restricted to precipitation at critical chain length, while for the pseudo-bulk models, radicals below the critical chain length also diffuse into and out of the microgels. Another example are the differences in descriptions of compartmentalization among the second and the third model. For better agreement, parameter values are adjusted. However, the adjusted values are mostly in the same order of magnitude. Hence, the similar values allow the same conclusions concerning their impact of the described mechanisms and confirm the general modeling framework.

Finally, by the sequential development of the three proposed models, the complexity of describing the microgel synthesis is resolved. New approaches such as the integration of quantum mechanical calculations can successfully support the model development to obtain profound insight into the microgel synthesis.

This thesis can serve as a foundation for future model-based investigation precipitation polymerization of microgels. Thereby, development and adjustment of the models is and should remain an ongoing process, targeted towards the specific application of the model. The most important directions for future development of the models and model-based investigations based upon the models are:

Extension of the model by diffusion limitation When comparing the three proposed models with different level of detail for the microgel synthesis, the transfer of model parameters from one model to the other lead to small inaccuracies. For better agreement with experimental data, the termination rate constants in the gel phase are adjusted. A first-principle description for their chain-length dependence of reaction termination could elucidate the impact of gelation on the reaction progress and make the need for parameter estimation obsolete.

Model-based development of advanced polymer structures The presented models not only provide deep insight into the microgel structure and improve the understanding of the process, they can also be employed for a prediction based on the process operation. In particular, due to the use of parameter values from quantum mechanical calculations incorporated into the process models, insight can be gained a priori and experimental effort should be reduced. While the benefits of microgel structures tailored to the applications are obvious, such an approach should be put to the test to synthesize microgels with constant radical co- or terpolymer compositions or core-shell microgels.

Towards first-principle predictive modeling The successful combination of quantum mechanical calculations and process modeling is successfully applied to describe the incorporation of cross-linker and even a more complex terpolymerization system. However, it could be an even more powerful tool when used to determine the parameter values for reaction mechanisms, which cannot be observed by measurements. Examples are the chain transfer to polymer, which contributes to long-branching polymers, and hence the formation of a polymer network without the addition of cross-linker, as well as the different termination rate constants for recombination and disproportionation, which cannot be distinguished in network formation. In combination with a description for chain-length dependence of termination, this could lead the way towards a first-principle predictive model.

Appendix A

Parameter values

Table A.1: Parameter values for chain propagation reaction from quantum mechanical calculation. Parameter values are calculated as described in Kröger et al. (2017). For liquid phase parameter values, infinite dilution assumption applies. For gel phase, dependence of propagation reaction on surrounding phase composition is calculated for a polymer weight fraction $w_P^g = 0.9$. The indices 1 and 2 represent the monomer VCL and the cross-linker BIS, respectively.

Rate constant	Liquid phase l k m^3 (mol s)$^{-1}$	Gel phase g k m^3 (mol s)$^{-1}$	ΔH_R kJ mol^{-1}
k_{p11}	$1.34 \cdot 10^0$	$7.167 \cdot 10^{-1}$	-83.2
k_{p12}	$3.60 \cdot 10^2$	$2.24 \cdot 10^2$	-87.4
k_{p21}	$3.43 \cdot 10^1$	$1.24 \cdot 10^1$	-74.8
k_{p22}	$8.10 \cdot 10^0$	$1.01 \cdot 10^0$	-77.8
k_{fm11}	$2.35 \cdot 10^{-2}$	$= k_{fm11}^l$	–
k_{fm12}	0	$= k_{fm12}^l$	–
k_{fm21}	$3.13 \cdot 10^{-3}$	$= k_{fm21}^l$	–
k_{fm22}	$2.49 \cdot 10^{-13}$	$= k_{fm22}^l$	–

Table A.2: Parameter values for chain propagation reaction from quantum mechanical calculation. Reaction rate constants are calculated from Arrhenius equation (Eq. (3.11)). Parameter values are calculated as described in Kröger et al. (2017). For liquid phase parameter values, infinite dilution assumption applies. For gel phase, dependence of propagation reaction on surrounding phase composition is calculated for a polymer weight fraction $w_{\mathrm{P}}^{\mathrm{g}} = 0.5$. The indices 1, 2, and 3 represent VCL, NIPAM and BIS, respectively.

	Liquid phase l		Gel phase g		
Rate constant	A	E_{A}	A	E_{A}	ΔH_{R}
	$\mathrm{m}^3\,(\mathrm{mol\ s})^{-1}$	$\mathrm{kJ\ mol}^{-1}$	$\mathrm{m}^3\,(\mathrm{mol\ s})^{-1}$	$\mathrm{kJ\ mol}^{-1}$	$\mathrm{kJ\ mol}^{-1}$
k_{p11}	$6.49 \cdot 10^4$	-30.81	$1.71 \cdot 10^4$	-29.42	-83.2
k_{p12}	$1.54 \cdot 10^5$	-23.53	$3.33 \cdot 10^4$	-21.25	-74.3
k_{p13}	$2.23 \cdot 10^5$	-18.39	$1.03 \cdot 10^5$	-18.35	-87.4
k_{p21}	$7.99 \cdot 10^4$	-13.86	$2.96 \cdot 10^4$	-12.75	-97.6
k_{p22}	$1.71 \cdot 10^5$	-16.16	$1.07 \cdot 10^5$	-17.30	-87.4
k_{p23}	$5.98 \cdot 10^5$	-18.17	$4.08 \cdot 10^5$	-18.90	-89.3
k_{p31}	$3.65 \cdot 10^4$	-19.91	$1.48 \cdot 10^4$	-20.01	-74.8
k_{p32}	$3.69 \cdot 10^5$	-30.05	$7.40 \cdot 10^4$	-27.91	-81.7
k_{p33}	$9.03 \cdot 10^4$	-26.63	$3.51 \cdot 10^4$	-28.31	-77.8

Table A.3: Parameter values for chain transfer reaction from quantum mechanical calculation. Reaction rate constants are calculated from Arrhenius equation (Eq. (3.11)). Parameter values are calculated as described by Kröger et al. (2017) for infinite dilution. Due to the low values for chain transfer in comparison to chain propagation, dependence of reaction constants on polymer weight fraction is neglected and the listed parameters are applied for liquid and gel phase. The indices 1, 2, and 3 represent VCL, NIPAM and BIS, respectively.

Rate constant	A	E_A	ΔH_R
	$\mathrm{m^3\ (mol\ s)^{-1}}$	$\mathrm{kJ\ mol^{-1}}$	$\mathrm{kJ\ mol^{-1}}$
k_{fm11}	$1.83 \cdot 10^8$	-65.0	40.39
k_{fm12}	-	-	-
k_{fm13}	0	-1	0
k_{fm21}	-	-	-
k_{fm22}	$4.64 \cdot 10^6$	77.6	-7.2
k_{fm23}	$4.37 \cdot 10^7$	120	70.63
k_{fm31}	$1.57 \cdot 10^9$	-76.8	34.37
k_{fm32}	$2.42 \cdot 10^6$	-87.2	6.38
k_{fm33}	$8.82 \cdot 10^6$	-128	83.45

Table A.4: General parameter values or calculation thereof.

Parameter	Value	Unit	Source
η_{BIS}	3		Assumption
η_{VCL}	15, 12, 10		Experiments (333 K, 343 K, 353 K)
μ^{l}	$e^{-3.7188+\frac{578.919}{-137.546+T}}$	Pa s	[2]
ρ_{PVCL}	$1.38\cdot10^{3}$	kg m^{-3}	Schneider et al. (2014)
ρ_{W}	$\frac{0.14395\cdot10^{-3}}{0.0112^{1+(1-T/649.727)^{0.05107}}}$	g m^{-3}	[2]
A_{d}	$9.19\cdot10^{14}$	s^{-1}	[1]
$C_{\mathrm{p,ins}}$	36.109	J K^{-1}	Mettler Toledo RTcal software
$c_{\mathrm{p,VCL}}$	2.3	J (g K)$^{-1}$	Calc. from Joback & Reid (1987) and Kwon et al. (2005)
$c_{\mathrm{p,W}}$	4.19	J (g K)$^{-1}$	Green & Perry (2008)
$D_{\mathrm{W,VCL}}$	$10^{-14.57}T/\mu^{\mathrm{l}}$	m^2 s^{-1}	Calc. from Wilke & Chang (1955)
$D_{\mathrm{W,BIS}}$	$10^{-14.55}T/\mu^{\mathrm{l}}$	m^2 s^{-1}	Calc. from Wilke & Chang (1955)
E_{A}	$1.24\cdot10^{5}$	kJ mol^{-1}	[1]
f_{I}	0.6		Assumption
k_{d}	$1.22\cdot10^{-4}$	s^{-1}	[1]
M_{BIS}	154.17	g mol^{-1}	
M_{NIPAM}	113.16	g mol^{-1}	
M_{VCL}	139.19	g mol^{-1}	
P_{BIS}	1.4846		QM, 90 w-% polymer content
P_{BIS}	$2.9671-0.0048T$		Linear correlation for parameter values calculated with QM, 50 w-% polymer, T = 333 K - 353 K
P_{I}	0.1		Assumption
P_{VCL}	2.8724		QM, 90 w-% polymer content
P_{VCL}	$8.9439-0.0177T$		Linear correlation for parameter values calculated with QM, 50 w-% polymer T = 333 K - 353 K
r_{min}, r_{max}	1e-10, 1e-7	m	
$w_{\mathrm{W}}^{\mathrm{g}}$	0.5		

[1] Wako Pure Chemical, Ltd. `https://www.wako-chemicals.de/en/product/v-50`, accessed 30.03.2017

[2] Dortmund Data Bank, 2018, `www.ddbst.com`

Table A.5: Estimated parameter values for copolymerization conversion model presented in Chapter 2. Partition coefficients P and efficiency factors f are estimated sequentially for pure VCL, pure BIS and VCL / BIS copolymerization.

Parameter	Unit	Value		
		VCL	BIS	VCL / BIS
		$i,j=1$	$i,j=2$	$i=1,j=2$
P_{I}		0.1	4.0	0.5
$f_{\mathrm{I}}^{\mathrm{g}}$		0.001	0.001	0.0013
$f_{\mathrm{I}}^{\mathrm{l}}$		0.8	0.99	0.986
$f_{\mathrm{PDB}i}^{\mathrm{g}}$			1	$1.2\cdot10^{-5}$
$k_{\mathrm{td}ij}^{\mathrm{l}}$	$\mathrm{m}^3\ (\mathrm{mol\ s})^{-1}$	$1.92\cdot10^{5}$	$1.05\cdot10^{7}$	$9.04\cdot10^{6}$
$k_{\mathrm{td}ij}^{\mathrm{g}}$	$\mathrm{m}^3\ (\mathrm{mol\ s})^{-1}$	$4.0\cdot10^{-2}$	$9.32\cdot10^{-4}$	$2.64\cdot10^{1}$

Table A.6: Estimated parameter values from simplified copolymerization model. The model described in Chapter C is used. While parameter values in Table C.1 are estimated sequentially, parameter values listed here are estimated simultaneously based on the same experiments. Also, the cross-termination rate constant is calculated by $k_{\mathrm{t}ij} = \left(k_{\mathrm{t}ii}k_{\mathrm{t}jj}\right)^{0.5}$. The cross-linking efficiency factor $f_{\mathrm{PDB}}^{\mathrm{g}}$ applies for both radicals of species VCL and BIS. Values are used in Chapter 3 (only for liquid phase) and in Chapter 4.

Parameter	Unit	Value	
		VCL	BIS
		$i=1$	$i=2$
$f_{\mathrm{PDB}}^{\mathrm{g}}$		$1.28\cdot10^{-3}$	
$k_{\mathrm{t}ii}^{\mathrm{l}}$	$\mathrm{m}^3\ (\mathrm{mol\ s})^{-1}$	$10^{5.22}$	$10^{6.37}$
$k_{\mathrm{t,diff}ii}^{\mathrm{g}}$	$\mathrm{m}^3\ (\mathrm{mol\ s})^{-1}$	$10^{-1.42}$	$10^{1.55}$

Appendix B

Two-phase precipitation polymerization model

Based on the reaction scheme in Table 2.1 and the listed assumptions in Chapter 2.2.1, the kinetic model described in the following is formulated. Initiator decomposition is described by the decomposition rate constant [1], which was also confirmed at reaction conditions by Raman spectroscopy. Mole balances for monomer and cross-linker are formulated, considering propagation and chain transfer to monomer reactions in both phases. Balances for radical and polymer are formulated for each phase separately. For primary radials, the quasi-steady state assumption (QSSA) is applied, which allows to derive the creation of radicals directly from the initiator decomposition. The initiator efficiency is accounted for by the factor $f_{\mathrm{I}}^{\mathrm{l,g}}$. The primary radical reacts with monomer or cross-linker proportionally to their respective fraction:

$$\frac{\mathrm{d}n_{\mathrm{I}}}{\mathrm{d}t} = -k_{\mathrm{d}}c_{\mathrm{I}}^{\mathrm{l}}V^{\mathrm{l}} - k_{\mathrm{d}}c_{\mathrm{I}}^{\mathrm{g}}V^{\mathrm{g}} \tag{B.1}$$

$$\frac{\mathrm{d}n_{\mathrm{M}_i}}{\mathrm{d}t} = -\sum_{q=\mathrm{l,g}} V^q\left(2f_{\mathrm{I}}^q k_{\mathrm{d}}c_{\mathrm{I}}^q f_{\mathrm{M}_i}^q + \sum_{j=1}^{2}(k_{\mathrm{p}ji}^q + k_{\mathrm{fm}ji}^q)c_{\mathrm{M}_i}^q c_{\lambda_0^j}^q\right) \tag{B.2}$$

$$f_{\mathrm{M}_2}^{\mathrm{l,g}} = c_{\mathrm{M}_2}^{\mathrm{l,g}}\left(c_{\mathrm{M}_1}^{\mathrm{l,g}} + c_{\mathrm{M}_2}^{\mathrm{l,g}}\right)^{-1} \tag{B.3}$$

$$f_{\mathrm{M}_1}^{\mathrm{l,g}} = 1 - f_{\mathrm{M}_2}^{\mathrm{l,g}} \tag{B.4}$$

$$c_o^{\mathrm{g}} = P_o c_o^{\mathrm{l}} \quad o \in \{\mathrm{I}, \mathrm{M}_1, \mathrm{M}_2\} \tag{B.5}$$

$$n_o^{\mathrm{l,g}} = V^{\mathrm{l,g}}c_o^{\mathrm{l,g}} \quad o \in \{\mathrm{I}, \mathrm{M}_1, \mathrm{M}_2\} \tag{B.6}$$

$$n_o = n_o^{\mathrm{l}} + n_o^{\mathrm{g}} \quad o \in \{\mathrm{I}, \mathrm{M}_1, \mathrm{M}_2\} \tag{B.7}$$

[1] Wako Pure Chemical, Ltd. https://www.wako-chemicals.de/en/product/v-50, accessed 30.03.2017

B.1 Liquid phase

The mole balances for dissolved oligomer are formulated for each specific chain length n up to $\eta - 1$ (Bunyakan et al. (1999)). This enables the direct derivation of the amount of substance of the oligomer with critical chain length and hence the precipitated polymer. For simplicity, the amount of substance of the radicals are summarized to the moment $n_{\lambda_0^i}$ where the index i indicates the type of the active (radical) unit of the polymer. Note that for the liquid phase, the moments of order $\Lambda = 0, 1$ are formulated for a finite number of chain units $\eta - 1$. The zeroth and first moment for inactive polymer is calculated independently of the polymer's chain length, considering chain transfer to monomer and termination by disproportionation. $\delta_{1,n}$ stands for the Kronecker Delta function with $\delta_{1,n} = 1$ for $n = 1$ and $\delta_{1,n} = 0$ otherwise.

$$\frac{\mathrm{d}n^{1}_{\mathrm{R}^i_n}}{\mathrm{d}t} = V^1\left(\left(2f^l_{\mathrm{I}}k_{\mathrm{d}}c^1_{\mathrm{I}}f^1_{\mathrm{p}} + \sum_{j=1}^{2} k^1_{\mathrm{fm}ji}c^1_{\lambda_0^j}c^1_{\mathrm{M}_i}\right)\delta_{1,n} + \sum_{j=1}^{2} k^1_{\mathrm{p}ji}c^1_{\mathrm{R}^j_{(n-1)}}c^1_{\mathrm{M}_i}\right.$$
$$\left. - \sum_{j=1}^{2}(k^1_{\mathrm{p}ij} + k^1_{\mathrm{fm}ij})c^1_{\mathrm{R}^i_n}c^1_{\mathrm{M}_j} - \sum_{j=1}^{2} k^1_{\mathrm{td}ij}c^1_{\mathrm{R}^i_n}c^1_{\lambda_0^j}\right) \tag{B.8}$$

$$\frac{\mathrm{d}n^1_{\mathrm{R}^i_\eta}}{\mathrm{d}t} = \sum_{j=1}^{2} k^1_{\mathrm{p}ji}c^1_{\mathrm{R}^j_{(\eta-1)}}c^1_{\mathrm{M}_i}V^1 \tag{B.9}$$

$$n^1_{\lambda^i_\Lambda} = \sum_{n=1}^{\eta-1} n^\Lambda n^1_{\mathrm{R}^i_n} \tag{B.10}$$

$$\frac{\mathrm{d}n^1_{\mu_\Lambda}}{\mathrm{d}t} = V^1\sum_{i=1}^{2}\sum_{j=1}^{2}\left(k^1_{\mathrm{fm}ij}c^1_{\lambda^i_\Lambda}c^1_{\mathrm{M}_j} + k^1_{\mathrm{td}ij}c^1_{\lambda^i_\Lambda}c_{\lambda^j_0}\right) \tag{B.11}$$

The reactions of radicals with a monomer terminal end and cross-linker terminal ends are expected to be equivalent ($k_{\mathrm{td}ij} = k_{\mathrm{td}ji}$) and hence the corresponding reaction is considered once only. For an observation of the degree of cross-linking, a balance for the pseudo-species PDB is introduced (Enright & Zhu (2000); Tobita & Hamielec (1989)). PDBs are formed in the continuous phase (n^1_{PDB}) by incorporation of cross-linker. The term $\dot{n}_{\mathrm{PDB},\eta}$ accounts for the PDBs which transfer from the liquid to the gel phase when a polymer chain precipitates.

$$\frac{\mathrm{d}n^1_{\mathrm{PDB}}}{\mathrm{d}t} = V^1\sum_{j=1}^{2}\left(k^1_{\mathrm{p}j2} + k^1_{\mathrm{fm}j2}\right)c^1_{\lambda_0^j}c^1_{\mathrm{M}_2} - \dot{n}_{\mathrm{PDB},\eta} \tag{B.12}$$

$$\dot{n}_{\mathrm{PDB},\eta} = \frac{n^1_{\mathrm{PDB}}}{n^1_{\lambda^1_0} + n^1_{\lambda^2_0} + n^1_{\mu_0}}\left(\dot{n}_{\mathrm{R}^1_\eta} + \dot{n}_{\mathrm{R}^2_\eta}\right) \tag{B.13}$$

B.2 Gel phase

For the gel phase, the moments of zeroth and first order are derived likewise to the liquid phase moments, but for an infinite number of monomer units. The moments are derived starting with oligomers with just one monomer unit since these are produced by initiation and chain transfer of monomer reactions. The collapsed oligomers contribute to the gel phase in terms of $\dot{n}_{\mathrm{R}^i_\eta}$, corresponding to their terminal unit of type i.

$$\frac{\mathrm{d}n^{\mathrm{g}}_{\lambda^1_0}}{\mathrm{d}t} = \dot{n}_{\mathrm{R}^1_\eta} + V^{\mathrm{g}}\Bigg(2f^{\mathrm{g}}_{\mathrm{I}}k_{\mathrm{d}}c^{\mathrm{g}}_{\mathrm{I}}(1-f^{\mathrm{g}}_{\mathrm{M}_2}) - \left(k^{\mathrm{g}}_{\mathrm{p}12}+k^{\mathrm{g}}_{\mathrm{fm}12}\right)c^{\mathrm{g}}_{\lambda^1_0}c^{\mathrm{g}}_{\mathrm{M}_2} + \left(k^{\mathrm{g}}_{\mathrm{p}21}+k^{\mathrm{g}}_{\mathrm{fm}21}\right)c^{\mathrm{g}}_{\lambda^2_0}c^{\mathrm{g}}_{\mathrm{M}_1} - f^{\mathrm{g}}_{\mathrm{PDB}1}k^{\mathrm{g}}_{\mathrm{p}12}c^{\mathrm{g}}_{\lambda^1_0}c^{\mathrm{g}}_{\mathrm{PDB}} - \sum_{j=1}^{2}k^{\mathrm{g}}_{\mathrm{td}1j}c^{\mathrm{g}}_{\lambda^1_0}c^{\mathrm{g}}_{\lambda^j_0}\Bigg) \tag{B.14}$$

$$\frac{\mathrm{d}n^{\mathrm{g}}_{\lambda^2_0}}{\mathrm{d}t} = \dot{n}_{\mathrm{R}^2_\eta} + V^{\mathrm{g}}\Bigg(2f^{\mathrm{g}}_{\mathrm{I}}k_{\mathrm{d}}c^{\mathrm{g}}_{\mathrm{I}}f^{\mathrm{g}}_{\mathrm{M}_2} - \left(k^{\mathrm{g}}_{\mathrm{p}21}+k^{\mathrm{g}}_{\mathrm{fm}21}\right)c^{\mathrm{g}}_{\lambda^2_0}c^{\mathrm{g}}_{\mathrm{M}_1} + \left(k^{\mathrm{g}}_{\mathrm{p}12}+k^{\mathrm{g}}_{\mathrm{fm}12}\right)c^{\mathrm{g}}_{\lambda^1_0}c^{\mathrm{g}}_{\mathrm{M}_2} + f^{\mathrm{g}}_{\mathrm{PDB}1}k^{\mathrm{g}}_{\mathrm{p}12}c^{\mathrm{g}}_{\lambda^1_0}c^{\mathrm{g}}_{\mathrm{PDB}} - \sum_{j=1}^{2}k^{\mathrm{g}}_{\mathrm{td}2j}c^{\mathrm{g}}_{\lambda^2_0}c^{\mathrm{g}}_{\lambda^j_0}\Bigg) \tag{B.15}$$

$$\begin{aligned}\frac{\mathrm{d}n^{\mathrm{g}}_{\lambda^1_1}}{\mathrm{d}t} = \eta\dot{n}_{\mathrm{R}^1_\eta} + V^{\mathrm{g}}\Bigg(&2f^{\mathrm{g}}_{\mathrm{I}}k_{\mathrm{d}}c^{\mathrm{g}}_{\mathrm{I}}(1-f^{\mathrm{g}}_{\mathrm{M}_2}) + k^{\mathrm{g}}_{\mathrm{p}11}c^{\mathrm{g}}_{\lambda^1_0}c^{\mathrm{g}}_{\mathrm{M}_1}\\ &+k^{\mathrm{g}}_{\mathrm{p}21}(c^{\mathrm{g}}_{\lambda^2_0}+c^{\mathrm{g}}_{\lambda^2_1})c^{\mathrm{g}}_{\mathrm{M}_1} - k^{\mathrm{g}}_{\mathrm{p}12}c^{\mathrm{g}}_{\lambda^1_1}c^{\mathrm{g}}_{\mathrm{M}_2} - f^{\mathrm{g}}_{\mathrm{PDB}1}k^{\mathrm{g}}_{\mathrm{p}12}c^{\mathrm{g}}_{\mathrm{PDB}}c^{\mathrm{g}}_{\lambda^1_1}\\ &+k^{\mathrm{g}}_{\mathrm{fm}11}(c^{\mathrm{g}}_{\lambda^1_0}-c^{\mathrm{g}}_{\lambda^1_1})c^{\mathrm{g}}_{\mathrm{M}_1} - k^{\mathrm{g}}_{\mathrm{fm}12}c^{\mathrm{g}}_{\lambda^1_1}c^{\mathrm{g}}_{\mathrm{M}_2} + k^{\mathrm{g}}_{\mathrm{fm}21}c^{\mathrm{g}}_{\lambda^2_0}c^{\mathrm{g}}_{\mathrm{M}_1} - \sum_{j=1}^{2}k^{\mathrm{g}}_{\mathrm{td}1j}c^{\mathrm{g}}_{\lambda^1_1}c^{\mathrm{g}}_{\lambda^j_0}\Bigg)\end{aligned} \tag{B.16}$$

$$\begin{aligned}\frac{\mathrm{d}n^{\mathrm{g}}_{\lambda^2_1}}{\mathrm{d}t} = \eta\dot{n}_{\mathrm{R}^2_\eta} + V^{\mathrm{g}}\Bigg(&2f^{\mathrm{g}}_{\mathrm{I}}k_{\mathrm{d}}c^{\mathrm{g}}_{\mathrm{I}}f^{\mathrm{g}}_{\mathrm{M}_2} + k^{\mathrm{g}}_{\mathrm{p}22}c^{\mathrm{g}}_{\lambda^2_0}c^{\mathrm{g}}_{\mathrm{M}_2} + k^{\mathrm{g}}_{\mathrm{p}12}(c^{\mathrm{g}}_{\lambda^1_0}+c^{\mathrm{g}}_{\lambda^1_1})c^{\mathrm{g}}_{\mathrm{M}_2}\\ &-k^{\mathrm{g}}_{\mathrm{p}21}c^{\mathrm{g}}_{\lambda^2_1}c^{\mathrm{g}}_{\mathrm{M}_1} + f^{\mathrm{g}}_{\mathrm{PDB}1}k^{\mathrm{g}}_{\mathrm{p}12}\left(c^{\mathrm{g}}_{\lambda^1_0}+c^{\mathrm{g}}_{\lambda^1_1}\right)c^{\mathrm{g}}_{\mathrm{PDB}} + f^{\mathrm{g}}_{\mathrm{PDB}2}k^{\mathrm{g}}_{\mathrm{p}22}c^{\mathrm{g}}_{\lambda^2_0}c^{\mathrm{g}}_{\mathrm{PDB}}\\ &+k^{\mathrm{g}}_{\mathrm{fm}22}(c^{\mathrm{g}}_{\lambda^2_0}-c^{\mathrm{g}}_{\lambda^2_1})c^{\mathrm{g}}_{\mathrm{M}_2} - k^{\mathrm{g}}_{\mathrm{fm}21}c^{\mathrm{g}}_{\lambda^2_1}c^{\mathrm{g}}_{\mathrm{M}_1} + k^{\mathrm{g}}_{\mathrm{fm}12}c^{\mathrm{g}}_{\lambda^1_0}c^{\mathrm{g}}_{\mathrm{M}_2} - \sum_{j=1}^{2}k^{\mathrm{g}}_{\mathrm{td}2j}c^{\mathrm{g}}_{\lambda^2_1}c^{\mathrm{g}}_{\lambda^j_0}\Bigg)\end{aligned} \tag{B.17}$$

$$\frac{\mathrm{d}n^{\mathrm{g}}_{\mu_\Lambda}}{\mathrm{d}t} = V^{\mathrm{g}}\sum_{i=1}^{2}\sum_{j=1}^{2}\left(k^{\mathrm{g}}_{\mathrm{fm}ij}c^{\mathrm{g}}_{\lambda^i_\Lambda}c^{\mathrm{g}}_{\mathrm{M}_j} + k^{\mathrm{g}}_{\mathrm{td}ij}c^{\mathrm{g}}_{\lambda^i_\Lambda}c^{\mathrm{g}}_{\lambda^j_0}\right)\quad \Lambda\in\{0,1\} \tag{B.18}$$

Further, balances for PDBs and cross-links X are formulated.

$$\frac{\mathrm{d}n^{\mathrm{g}}_{\mathrm{PDB}}}{\mathrm{d}t} = \dot{n}_{\mathrm{PDB},\eta} + V^{\mathrm{g}}\left(\sum_{j=1}^{2}\left((k^{\mathrm{g}}_{\mathrm{p}j2}+k^{\mathrm{g}}_{\mathrm{fm}j2})c^{\mathrm{g}}_{\lambda^j_0}c^{\mathrm{g}}_{\mathrm{M}_2} - f^{\mathrm{g}}_{\mathrm{PDB}j}k^{\mathrm{g}}_{\mathrm{p}j2}c^{\mathrm{g}}_{\lambda^j_0}c^{\mathrm{g}}_{\mathrm{PDB}}\right)\right) \tag{B.19}$$

$$\frac{\mathrm{d}n_{\mathrm{X}}^{\mathrm{g}}}{\mathrm{d}t} = V^{\mathrm{g}} \sum_{j=1}^{2} f_{\mathrm{PDB}_j}^{g} k_{\mathrm{p}j2}^{\mathrm{g}} c_{\lambda_0^j}^{\mathrm{g}} c_{\mathrm{PDB}}^{\mathrm{g}} \tag{B.20}$$

The number-average molecular weight M_{n} of the primary chain polymers in the gel particles is derived from the zeroth moment and first moments without the cross-links, as these would be counted double, otherwise. The average molar mass of repeating unit in a polymer chain M_{unit} is derived from the molar mass of the monomers M_{M_1} and the cross-linker M_{M_2} and the amount of reacted monomer and cross-linker ($n_{\mathrm{M}_{\mathrm{reac1}}}, n_{\mathrm{M}_{\mathrm{reac2}}}$), respectively. The total polymer produced in both phases n_{P} equals the sum of the moments of zeroth order.

$$l_{\mathrm{avg}} = \frac{n_{\mu_1}^{\mathrm{g}} + n_{\lambda_1^1}^{\mathrm{g}} + n_{\lambda_1^2}^{\mathrm{g}}}{n_{\mu_0}^{\mathrm{g}} + n_{\lambda_0^1}^{\mathrm{g}} + n_{\lambda_0^2}^{\mathrm{g}}} \tag{B.21}$$

$$M_{\mathrm{n}} = l_{\mathrm{avg}} M_{\mathrm{unit}} - n_{\mathrm{X}} M_{\mathrm{M}_2} \tag{B.22}$$

$$M_{\mathrm{unit}} = \frac{M_{\mathrm{M}_1} n_{\mathrm{M}_{\mathrm{reac1}}} + M_{\mathrm{M}_2} n_{\mathrm{M}_{\mathrm{reac2}}}}{n_{\mathrm{M}_{\mathrm{reac1}}} + n_{\mathrm{M}_{\mathrm{reac2}}}} \tag{B.23}$$

$$n_{\mathrm{P}} = n_{\lambda_0^1}^{1} + n_{\lambda_0^2}^{1} + n_{\mu_0}^{1} + n_{\lambda_0^1}^{\mathrm{g}} + n_{\lambda_0^2}^{\mathrm{g}} + n_{\mu_0}^{\mathrm{g}} \tag{B.24}$$

For comparison with Raman measurements, mass fractions are calculated for monomer w_{M_1} and polymer w_{P} with the total mass m_{tot}. w_{P} is computed from the closure condition. Validation with $w_{\mathrm{P}}^{\mathrm{val}} = \frac{n_{\mathrm{P}}^{\mathrm{g}} M_{\mathrm{n}}}{m_{\mathrm{tot}}}$ shows an error of $< 1\,\%$.

$$w_o = \frac{m_o}{m_{\mathrm{tot}}} \quad o \in \{\mathrm{I}, \mathrm{M}_1, \mathrm{M}_2, \mathrm{S}, \mathrm{W}\} \tag{B.25}$$

$$w_{\mathrm{P}} = 1 - w_{\mathrm{I}} - w_{\mathrm{M}_1} - w_{\mathrm{M}_2} - w_{\mathrm{S}} - w_{\mathrm{W}} \tag{B.26}$$

B.3 Polymer volume

From the precipitation mechanism, the mass of the gel phase can be derived. But the calculation of the concentrations in the particular phases requires the determination of volumes V^{l} and V^{g}. As the experimental separation and density determination of the gel phase at synthesis conditions is difficult and error-prone, an inverse approach was pursued.

In Maldonado-Parra (2019), the volume of polymer chains in water is derived from calculations using PC-SAFT. The thermodynamic modeling using PC-SAFT equations of state is published in Arndt & Sadowski (2012) for the example of PNIPAM hydrogels. The volume of polymer chains V_{chain} for 20 to 1000 repeating units l_{avg} is calculated for varying reactor temperatures $T = 333\,K - 353\,K$ and cross-linker to monomer mass ratios $y_{\mathrm{cross}} = 0.03 - 0.05$.

The calculated volumes are correlated to temperature and cross-linker content, resulting in Eq. (B.27) with fitted parameters ($A = 8.15 \cdot 10^{-29}\,\mathrm{m}^3$, $F_1 = 4.01 \cdot 10^{-29}\,\mathrm{m}^3$, $F_2 = -0.367$, $G = 2.49 \cdot 10^{-31}\,\mathrm{m}^3\mathrm{K}^{-1}$, $H = 2.86 \cdot 10^{-28}\,\mathrm{m}^3$). V^{g} is calculated from the sum of polymer molecules (Avogadro constant N_{A}) and their average volume V_{chain} (Eq. (B.28)). V^{l} is calculated from the constant total volume V_{tot} (Eq. (B.29)). Further, the gel volume fraction in percent $\phi^{\mathrm{g}} = V^{\mathrm{g}}/V_{\mathrm{tot}} \cdot 100\%$ can be calculated.

$$V_{\mathrm{chain}} = l_{\mathrm{avg}}(A + \frac{F_1 l_{\mathrm{avg}}}{F_2 + l_{\mathrm{avg}}} + GT + Hy_{\mathrm{cross}}) \tag{B.27}$$

$$V^{\mathrm{g}} = V_{\mathrm{chain}}(n^{\mathrm{g}}_{\lambda_0^1} + n^{\mathrm{g}}_{\lambda_0^1} + n^{\mathrm{g}}_{\mu_0})N_{\mathrm{A}} \tag{B.28}$$

$$V_{\mathrm{tot}} = V^{\mathrm{l}} + V^{\mathrm{g}} \tag{B.29}$$

Without sufficient knowledge of the growth mechanism of aggregation, the growth of a single particle can not be simulated. Hence, the particle volume V^{p} is calculated subsequently. A possible explanation for a monodisperse distribution of the final particles is, that particles are formed in a short period after initiation only (Duracher et al. (1999)). Then, the number of particles remains constant and particles grow likewise. The absolute number of particles N^{p} is calculated from the final gel volume V^{g} and final particle volumes $V^{\mathrm{p}}_{\mathrm{end}}$ in collapsed state, derived from DLS measurements at 323 K (above the VPTT). As the gel volume is simulated at any time of polymerization, the volume of a single average particle can be approximated throughout the reaction.

This and Eq. (B.28) entail the error, that a particle is formulated as an assembly of polymer chains with the length of l_{avg} ($V^{\mathrm{p}} = \sum V_{\mathrm{chain}}(l_{\mathrm{avg}})$) instead of a particle as one macromolecule of cross-linked polymer chains ($V^{\mathrm{p}} = V_{\mathrm{chain}}(\sum l_{\mathrm{avg}})$). The comparison of the final particle volume as calculated from one molecule with a high monomer units number and calculation of the equivalent in monomer units as composite particle with small average chain length revealed an sufficiently small error of $< 0.5\,\%$. The calculated density for the final gel phase ($1220 - 1330\,\mathrm{kg\,m}^{-3}$) is similar to the experimentally determined density of pure PVCL (without water) ($1380 \pm 40\,\mathrm{kg\,m}^{-3}$ Schneider et al. (2014)).

B.4 Heat loss coefficients

Table B.1: Parameter values for heat loss through reactor lid, determined from reference experiment. Two values were determined for 343 K, once during heating ([h]) and once while cooling ([c]). $\alpha = kA_{\text{lid}}$ also includes the heat transfer of the gaseous phase in the reactor above the liquid, which is equally heated and cooled, respectively and hence, for heat loss, steady state cannot be obtained. The used heat loss coefficients α approximate steady state when a constant reactor temperature is observed.

Temperature T	Heat loss coefficient α
K	W K^{-1}
313	0.0314
333	0.0451
343[h]	0.0477
353	0.0621
363	0.0761
343[c]	0.0272

B.5 Simulation and calorimetry data for copolymerization: Direct comparison

In Figure B.1, experiments and simulation are compared directly in terms of Σ_{R} for the copolymerization reactions. Experimental Σ_{R} is calculated with the parameters used in the model (cf. Appendix Table A.1 and Table A.4) from the energy balance. The bars depict the standard deviation of the three repetitions of the experiment and for illustration, the measurement values are shown in 10 s intervals only.

For the lowest cross-linker concentration 1.2 mol-%, the best agreement of simulation and experiment is obtained. For 5.0 mol-%, the difference between simulation and experiment is most distinct. The graphs show the large standard deviations of the measurements at the beginning of the polymerization. Within the first 50 s (1.2 mol-%) to 100 s (5.0 mol-%), which is the time of cross-linker consumption, the standard deviation are most pronounced, expressing the highest deviations among the experiment repetitions.

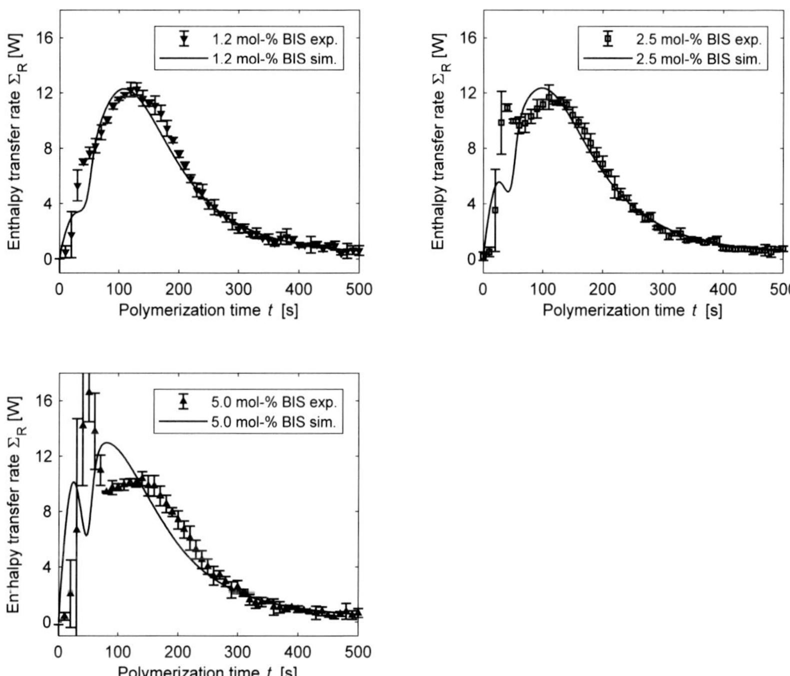

Figure B.1: Simulated and experimental Σ_R for experiments with 1.2, 2.5 and 5.0 mol-% BIS. For improved readability, measurements are illustrated in 10 s intervals only. Standard deviation is calculated from three repetitions of the experiments.

Appendix C

Case study: Precipitation terpolymerization for cross-linked PVCL-PNIPAM-based microgels

Likewise to PVCL-based microgels, poly(*N*-isopropylacrylamide) (PNIPAM)-based microgels as well as PVCL-PNIPAM-copolymer-based microgels can be synthesized by precipitation polymerization (Pich & Richtering (2011)). Balaceanu et al. (2013) reported for microgels from PVCL and PNIPAM uniform internal structures, although NIPAM is the more reactive monomer. To help understand the driving forces of the microgel formation while also illustrating the applicability of parameters from quantum mechanical (QM) calculation to different systems, the approach presented in Chapter 2 is applied to the terpolymerization precipitation of PVCL-PNIPAM-based microgels with the cross-linker BIS. The model is fitted to previously presented and new experimental data and applied for a rough prediction of the microgel's internal structure.

C.1 Modeling of precipitation terpolymerization

In Chapter 2, the synthesis of PVCL-based microgel by precipitation copolymerization model is abstracted into two phases. Mass balances for all reaction species are formulated, whereas the primary chains of polymers are balanced applying the terminal model and the method of moments. Mass transfer between the phases is described for small molecules by phase equilibrium with distribution coefficients P and for polymer by precipitation of the radicals at a finite critical chain length η. The key feature of this approach is that the high number of unknown parameter values is reduced by the use of reaction rate constants and distribution coefficients calculated from transition state theory and COSMO-RS. In this case study, the approach is applied to a model for the terpolymerization of VCL (1) and NIPAM (2) with the cross-linker BIS (3).

Few alterations are applied to the model in order to reduce the number of parameter values to be estimated and decrease the model complexity without loss of information. These alterations are:

1. Constant initiator efficiencies $f_{\mathrm{I}} = 0.6$ in liquid and gel phase, which is a mean efficiency for azo-initiators.

2. Cross-linking in the liquid phase is neglected ($f^{\mathrm{l}}_{\mathrm{PDB}} = 0$) while in the gel phase, cross-linking reactions are assumed to be equally limited by the formation of the polymer network, resulting in one cross-linking efficiency factor $f^{\mathrm{g}}_{\mathrm{PDB}}$.

3. Termination rate constants $k^{\mathrm{g}}_{\mathrm{t}}$ in the gel phase represent the apparent rate constants which account for diffusion limitation ($k^{\mathrm{g}}_{\mathrm{t,diff}}$). They are linked to the termination rate constants of the liquid phase by $k^{\mathrm{g}}_{\mathrm{t}} = (k^{\mathrm{l}}_{\mathrm{t}}\, k^{\mathrm{g}}_{\mathrm{t,diff}})/(k^{\mathrm{l}}_{\mathrm{t}} + k^{\mathrm{g}}_{\mathrm{t,diff}})$ (Buback et al. (1992)).

4. The gel phase volume, V^{g}, is calculated from the pure polymer densities ρ and 50 w-% water. The volume of the liquid phase, V^{l} follows from $V_{\mathrm{tot}} = V^{\mathrm{l}} + V^{\mathrm{g}} = \mathrm{const.}$.

5. The liquid phase serves as initiation phase, due to a small critical chain length and $P_{\mathrm{M}} > 1$. To calculate the mass flow by precipitation, first the pseudo-homopolymerization approach (Storti et al. (1989)) and then the quasi-steady state assumption (QSSA) for radicals are applied. Thus, the $3(\eta - 1)$ differential equations for three comonomers are replaced by $3 + 1$ algebraic equations (cf Appendix D.1).

C.1.1 Quantum mechanical calculations

The chain propagation rate constants k_{p} with the corresponding reaction enthalpies ΔH_{R} are calculated by transition state theory depending on the composition of both phases: For the liquid phase infinite dilution and for the gel phase a composition of 50 w-% polymer and 50 w-% water is assumed. The corresponding parameter values are used as listed in Table A.2 and Table A.3, which were derived as described in Kröger et al. (2017). VCL and NIPAM cross-propagation rate constants are calculated accordingly for the liquid phase ($k^{\mathrm{l}}_{\mathrm{p12}} = 47.3\ \mathrm{m^3(mol\ s)^{-1}}$, $k^{\mathrm{l}}_{\mathrm{p21}} = 620.0\ \mathrm{m^3(mol\ s)^{-1}}$) and the gel phase ($k^{\mathrm{g}}_{\mathrm{p12}} =$

$19.6\,\text{m}^3(\text{mol s})^{-1}$, $k^{\text{g}}_{\text{p21}} = 315.7\,\text{m}^3(\text{mol s})^{-1}$). The unknown distribution coefficient for NIPAM is calculated with COSMO-RS ($P_{\text{M}_2} = 2.52$).

C.1.2 Experiments

All experiments are performed in a Mettler Toledo RTcal 0.5 L real-time calorimeter in isothermal control mode at 70 °C and ambient pressure. The reactor is filled with $3 \cdot 10^{-4}\,\text{m}^3$ water and all reaction species, monomers (homo- and copolymerization: $0.106\,\text{mol}\,\text{L}^{-1}$ VCL or NIPAM; terpolymerization: $0.053\,\text{mol}\,\text{L}^{-1}$ VCL and NIPAM) with or without cross-linker BIS (2.5 mol-% of monomer) are dissolved as well as the stabilizer Cetyltrimethylammoniumbromide (CTAB, 1 w-% of monomer). Before the reaction, the reactor is purged with nitrogen. 0.1 g thermal initiator AMPA is added. The experiment runs for 1.5-2 h after initiation before it is terminated. Calorimetry and Raman spectroscopy are performed simultaneously. VCL and BIS experiments are used as presented in Chapter 2. In-line Raman spectra are evaluated with Indirect Hard Modeling (IHM) according to the methods described in Meyer-Kirschner et al. (2016) and Meyer-Kirschner et al. (2018) to determine monomer mass fractions w_{M_i} of VCL and NIPAM. Despite purging, the experimental data reveal an inhibition phase right after initiation, which is removed for parameter estimation and comparison with simulation data.

The remaining unknown parameter values for initiator partition coefficient P_{I}, termination rate constants k^{l}_{t}, $k^{\text{g}}_{\text{t,diff}}$, and cross-linking efficiency f_{PDB} are determined by parameter estimation. The experimental data from Raman spectroscopy and the model are linked by the weight fraction balance for the monomers VCL and NIPAM:

$$\frac{\mathrm{d}w_{\text{M}_i}}{\mathrm{d}t} = \frac{M_{\text{M}_i}}{m_{\text{tot}}}\Big(-V^{\text{l}} \sum_{j=1}^{3} (k^{\text{l}}_{\text{p}ji} + k^{\text{l}}_{\text{fm}ji}) c^{\text{l}}_{\lambda_0^j} c^{\text{l}}_{\text{M}_i} - V^{\text{g}} \sum_{j=1}^{3} (k^{\text{g}}_{\text{p}ji} + k^{\text{g}}_{\text{fm}ji}) c^{\text{g}}_{\lambda_0^j} c^{\text{g}}_{\text{M}_i} \Big). \tag{C.1}$$

Thereby, the termination rate constants only indirectly affect w_{M_i} through the radical concentrations $c_{\lambda_0^j}$. In addition, reaction calorimetry provides information on the overall polymerization progress. The enthalpy transfer rate of the reaction, Σ_{R}, represents the heat released at constant temperature by the exothermic propagation reactions of monomer and cross-linker:

$$\Sigma_{\text{R}} = \sum_{q=\text{l,g}} \sum_{i=1}^{3} \left(\sum_{j=1}^{3} \Big(-\Delta H_{\text{R}ij} V^q k^q_{\text{p}ij} c^q_{\lambda_0^i} c^q_{\text{M}_j} \Big) - \Delta H_{\text{R}i3} V^q f^q_{\text{PDB}i} k^q_{\text{p}i3} c^q_{\lambda_0^i} c^q_{\text{PDB}} \right). \tag{C.2}$$

It also accounts for the consumption of BIS and reactions of the PDBs, while the BIS concentration is below the detection limit of the utilized Raman spectrometer and therefore cannot be measured directly. Thus, both measurement techniques complement each other.

C.2 Results and discussion

Modeling and parameter estimation with the maximum likelihood method are both performed with gProms version 4.2.0 [1] on a 64-bit Intel Core i3-3225 CPU with 3.3 GHz running Windows 7. The parameter values are estimated sequentially and transferred, beginning with homopolymerizations of VCL and NIPAM, followed by their copolymerizations with BIS, respectively, and last terpolymerization of VCL, NIPAM and BIS.

C.2.1 Homo- and copolymerization of VCL and NIPAM

The model with parameter values from QM calculation is evaluated for polymerization reactions of pure VCL and NIPAM. Figure C.1 shows the enthalpy transfer rate Σ_R calculated from measurements in comparison to the results with the fitted model. Σ_R is plotted over time, whereas $t = 0\,\mathrm{s}$ corresponds to the moment of initiation in simulation and for experiments is located after the observed inhibition phase. Both curves, VCL and NIPAM, show a good agreement of the fitted model with experimental data.

The experimental data shows that the overall polymerization of VCL proceeds faster than of NIPAM, as Σ_R for VCL increases faster after initiation and obtains its maximum after 100 s while the curve for NIPAM obtains its significantly lower maximum after 150 s. This is counter-intuitive, as the calculated k_{p22} (Kröger et al. (2017)) indicate a faster propagation rate for NIPAM. Consequently, the estimated k_t to fit the experimental data differ significantly. A low estimated value $k^g_{t,diff11} = 3.54 \cdot 10^{-2}\,\mathrm{m}^3(\mathrm{mol\ s})^{-1}$ for VCL dominates k^g_{t11}, expressing a significant diffusion limitation of the termination reaction. This accelerates the overall polymerization reaction. For NIPAM, the higher estimated $k^g_{t,diff22} = 9.89 \cdot 10^3\,\mathrm{m}^3(\mathrm{mol\ s})^{-1}$ suggests that the termination reaction is less affected and the overall polymerization reaction proceeds slower.

Figure C.2 depicts the results of the fitted model in comparison to experimental data for copolymerizations with the same amount of cross-linker. For both reaction systems, the parameter values from QM calculation can be employed to calculate the reaction progress, as the experimental data can be fitted with good agreement. For VCL-BIS, the first maximum in the Σ_R measurement results from the faster consumption of cross-linker, represented by higher cross-propagation reaction rate constants, as discussed in Chapter 2. Note that the improved fit for VCL-BIS results from a reevaluation of the experimental data regarding the inhibition phase. Σ_R for NIPAM-BIS on the other hand does not indicate such inhomogeneous consumption.

The estimated $k^g_{t,diff}$ for both reactions differ significantly. A good fit for VCL-BIS is only

[1]Process Systems Enterprise, gProms, www.psenterprise.com/gproms 1997-2020

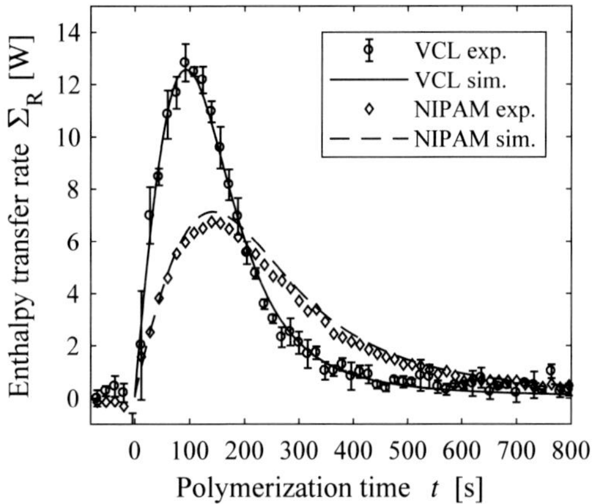

Figure C.1: Comparison of experiments and fitted model for homopolymerization of VCL in terms of Σ_R.

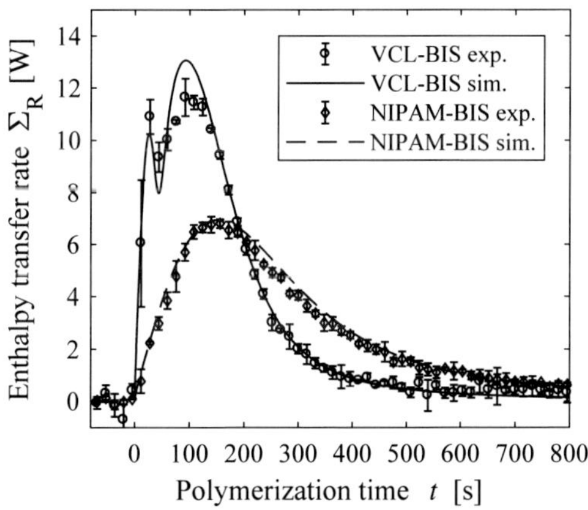

Figure C.2: Comparison of experiments and fitted model for copolymerizations with BIS in terms of Σ_R.

obtained, when individual diffusion limitations for the termination reactions are introduced ($k^{g}_{t,diff11} < k^{g}_{t,diff13} < k^{g}_{t,diff33}$). All estimated parameter values are listed in Table C.1 For NIPAM-BIS, a constant $k^{g*}_{t,diff2}$ for the different termination reactions is sufficient, but it needs to be determined for reactions with different cross-linker concentrations individually. Then, $k^{g*}_{t,diff2}$ decreases with increasing cross-linker concentration, causing the overall polymerization to proceed faster. Thus, for the system of NIPAM-BIS, diffusion limitation depends on the cross-linker concentration, whereas for the VCL-BIS system, the monomer VCL itself, e.g., by steric limitation of the molecular structure, causes a limitation of the termination reaction.

Table C.1: Termination rate constants estimated with the terpolymerization model. The estimated cross-link efficiency factor is f_{PDB} = 0.021. [a] estimated for homopolymerization, [b] estimated for copolymerization, [c] estimated for terpolymerization.

Rate constant	Liquid phase [1] $m^3\ (mol\ s)^{-1}$	Diffusion term $^{g}_{diff}$ $m^3\ (mol\ s)^{-1}$
$k_{t,11}$	$10^{5.23}$	$10^{-1.45}$
$k_{t,12}$	$10^{3.98}$	$10^{2.35}$
$k_{t,13}$	$10^{3.11}$	$10^{0.48}$
$k_{t,22}$	$10^{11.41}$	$10^{3.99}$ [a], $10^{2.76}$ [b], $10^{2.35}$ [c]
$k_{t,23}$	$10^{10.22}$	$= k_{t,22}$
$k_{t,33}$	10^{7}	$= k_{t,22}$

C.2.2 Terpolymerization

The model is applied to the terpolymerization system. Thereby, the observations from copolymerization are incorporated in terms of the fitted parameter values: while $k^{g}_{t,diff11}$ and $k^{g}_{t,diff13}$ are transferred as previously estimated, the $k^{g*}_{t,diff2}$ is estimated again due to the different composition.

Figure C.3 shows good agreement between Σ_R determined from experiments and the fitted model. The maximum enthalpy transfer rate is significantly higher (> 18 W) than for the copolymerizations and the overall reaction proceeds fast. For the fitted model, this results from high k_{p12} and k_{p21} in combination with low estimates for $k^{g}_{t,diff12}$ and $k^{g*}_{t,diff2}$. Hence, the overall reaction is accelerated.

The comparison of the measurements of the mass fractions w_M of VCL and NIPAM with the fitted model is illustrated in Figure C.4. NIPAM is consumed slightly faster than VCL. Still, high VCL cross-propagation rate constants and the preferential accumulation of VCL in

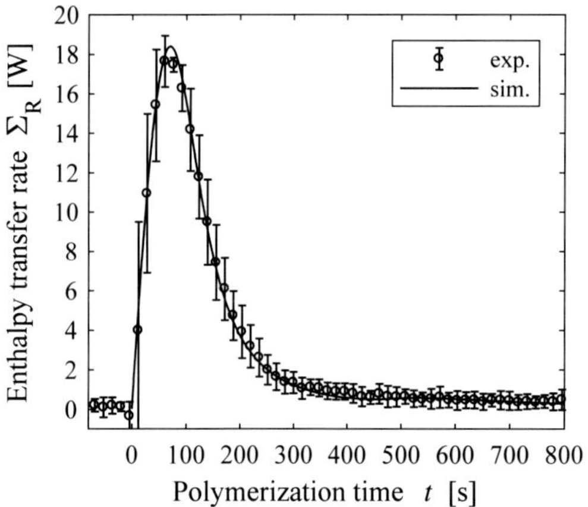

Figure C.3: Comparison of experiments and simulation with fitted model for terpolymerization in terms of enthalpy transfer rate from calorimetry. Time $t = 0$ s denotes the initiation.

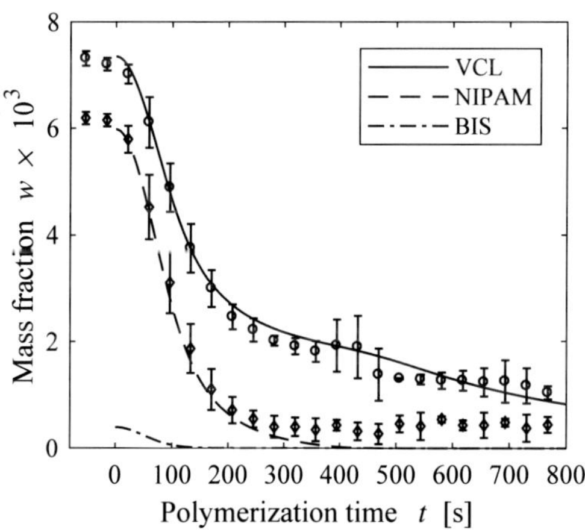

Figure C.4: Comparison of experiments and simulation with fitted model for terpolymerization in terms of monomer mass fractions from Raman spectroscopy. BIS concentration is below the detection limit and hence cannot be determined from in-line spectra. Time $t = 0$ s denotes the initiation.

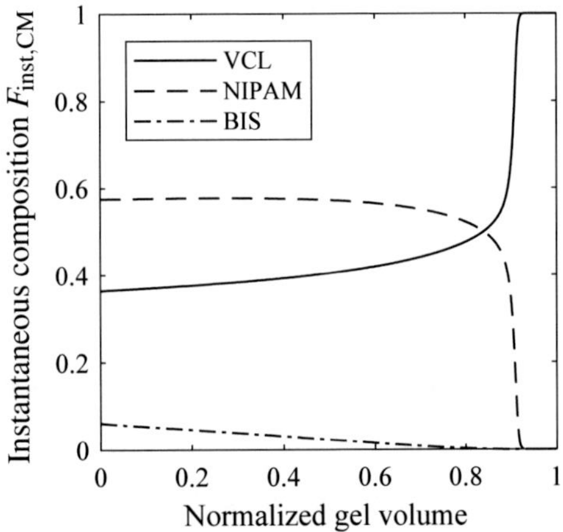

Figure C.5: Instantaneous comonomer mole fraction compared to the instantaneous normalized radius.

the gel phase ($P_{M_1} > P_{M_2}$) promote the consumption of VCL in the presence of NIPAM and BIS. After 300 s, w_{M_2} is below $1 \cdot 10^{-3}$ and thereby in the size range of the prediction error of the chemometric model (Meyer-Kirschner et al. (2018)). With the combination of both measurements, the cross-linker consumption can be predicted with the fitted model.

The combination of comonomer consumption and the predicted gel phase growth provides a rough prediction of the internal microgel structure, as illustrated in Figure C.5. The instantaneous copolymer fraction $F_{inst,CM}$ represents the reaction rate fractions of the comonomers in the gel phase and V^g is predicted from the mass transfer due to phase equilibrium and precipitation. The instantaneous BIS mole fraction declines over the radius. As the cross-linker is incorporated in the polymer network, the instantaneous radius represents the maximum range of its position in the microgel, indicating a higher cross-linked core. The instantaneous mole fractions of VCL and NIPAM are mostly constant over the radius, though the faster NIPAM consumption results in a higher NIPAM fraction. For the normalized radius close to unity, VCL is the only monomer left. Hence, the final microgels have a core with radially uniform composition of VCL and NIPAM and a thin outer corona of pure PVCL. The prediction of the microgel structure agrees with the uniform distribution of VCL and NIPAM described by Balaceanu et al. (2013).

C.3 Conclusion

In this case study, the precipitation polymerization model for VCL-based microgels presented in Chapter 2 is extended to the synthesis of microgels based on PVCL and PNIPAM. Parameter values from QM calculation are employed to reduce the number of unknown parameter values. The unknown termination rate constants are estimated from experimental data for increasing system complexities. Thereby, substantial differences in the termination rate constants are observed for homo- and copolymerizations of VCL and NIPAM: A lower termination rate for VCL causes the overall reaction to proceed faster, despite its lower propagation rate constants. For terpolymerization, the overall polymerization proceeds faster due to high propagation reaction rates and low termination rate constants. Thereby, NIPAM is consumed slightly faster due to its higher propagation rate constants. In agreement with previous experimental observations, the prediction of the internal microgel structure shows a radially uniform distribution of the comonomers VCL and NIPAM.

Appendix D

Model Equations for Pseudo-Bulk Model and additional information

D.1 Liquid phase modeling

For the concentration of radicals in the liquid phase, the QSSA is applied. Based on the reactions listed in Table 3.1 and applying the method of moments, the equation for the concentration of radicals $c^{\mathrm{l}}_{\lambda_0}$, corresponding to the zeroth order moment,

$$2f^{\mathrm{l}}k_{\mathrm{d}}c^{\mathrm{l}}_{\mathrm{I}} + \mathcal{R}_{\mathrm{des}} = \left(k^*_{\mathrm{p}} + k^*_{\mathrm{PDB}}\right) c^{\mathrm{l}}_{\mathrm{R}_\eta} + k^*_{\mathrm{t}}(c^{\mathrm{l}}_{\lambda_0})^2 + \mathcal{R}_{\mathrm{e}}. \tag{D.1}$$

is obtained. Eq. (D.1) balances the radicals entering the liquid phase by initiation or desorption from particles on the L.H.S. with the radicals leaving the liquid phase by precipitation, termination and entry into particles on the R.H.S.. Therein, f^{l}, k_{d} and $c^{\mathrm{g}}_{\mathrm{I}}$ are the initiator efficiency, initiator decomposition rate and initiator concentration in the liquid phase, $\mathcal{R}_{\mathrm{des}}$ and $\mathcal{R}_{\mathrm{e}}$ are radical desorption and entry rates and k^*_{p}, k^*_{PDB} and k^*_{t} are the weighted propagation, cross-linking and termination rates. For initiator decomposition and initiation reaction, the QSSA is applied.

Likewise, the first order moment provides the concentration of build-in monomer units in radicals:

$$2f^{\mathrm{l}}k_{\mathrm{d}}c^{\mathrm{l}}_{\mathrm{I}} + \left(k^*_{\mathrm{p}} + k^*_{\mathrm{fm}}\right) c^{\mathrm{l}}_{\lambda_0} + \mathcal{R}_{\mathrm{des}} = \left(k^*_{\mathrm{p}} + k^*_{\mathrm{PDB}}\right) \eta c^{\mathrm{l}}_{\mathrm{R}_\eta} + \left(k^*_{\mathrm{fm}} + k^*_{\mathrm{t}} c^{\mathrm{l}}_{\lambda_0}\right) c^{\mathrm{l}}_{\lambda_1} + \mathcal{R}_{\mathrm{e}} l_{\mathrm{avg}}, \tag{D.2}$$

where k^*_{fm} is the weighted chain transfer to monomer rate and l_{avg} is the average number of monomer units per radical, calculated from

$$l_{\text{avg}} = \frac{c_{\lambda_1}}{c_{\lambda_0}}. \tag{D.3}$$

The copolymerization mechanism is therein captured by the pseudo-homopolymerization approach (Storti et al. (1989)). With the Mayo-Lewis equation for multiple monomers

$$\sum_{j=1}^{2} k_{\text{p}ij} p_i c_{\text{M}_j} = \sum_{j=1}^{2} k_{\text{p}ji} p_j c_{\text{M}_i} \quad i = 1 \tag{D.4}$$

$$\sum_{i=1}^{2} p_i = 1. \tag{D.5}$$

the fractions p_i^1 and p_j^1 of radicals with terminal end of species i and j are calculated, respectively. The weighted reaction rates in Eqs. (D.1)-(D.2) are calculated according to

$$k_{\text{p}}^* = \sum_{i,j=1}^{2} k_{\text{p}ij} p_i c_{\text{M}_j}, \tag{D.6}$$

$$k_{\text{PDB}}^* = \sum_{i=1}^{2} f_{\text{PDB}} k_{\text{p}i2} p_i c_{\text{M}_2}, \tag{D.7}$$

$$k_{\text{fm}}^* = \sum_{i,j=1}^{2} k_{\text{fm}ij} p_i c_{\text{M}_j}, \tag{D.8}$$

$$k_{\text{t}}^* = \sum_{i,j=1}^{2} k_{\text{t}ij} p_i p_j. \tag{D.9}$$

The formulation above with the QSSA and the pseudo-homopolymerization approach is a simplified form of the comprehensive formulation described in Appendix B.1 and previously mentioned in Appendix C. The simplified form is used in this chapter to reduce the number of differential equations.

The radical entry and desorption rates in Eqs. (D.1)-(D.2) are calculated from the radical entry and desorption rates into the particles:

$$\mathcal{R}_{\text{e}} = \int_{r_{\min}}^{r_{\max}} \mathcal{R}_{\text{e}}^{\text{p}}(r) \mathcal{F}(r,t) \text{d}r, \tag{D.10}$$

$$\mathcal{R}_{\text{des}} = \frac{V}{V^1} \int_{r_{\min}}^{r_{\max}} \Sigma_{i=1}^{2} \mathcal{R}_{\text{des}i}^{\text{p}}(r) \mathcal{F}(r,t) \text{d}r. \tag{D.11}$$

Assuming steady state for radicals for any chain length l up to the critical chain length η, the following expression for the concentration $c^{\mathrm{l}}_{\mathrm{R}_\eta}$ of radicals with critical chain length is derived:

$$c^{\mathrm{l}}_{\mathrm{R}_\eta} = \frac{\left(2f^{\mathrm{l}}k_{\mathrm{d}}c^{\mathrm{l}}_{\mathrm{I}} + k^*_{\mathrm{fm}}c^{\mathrm{l}}_{\lambda_0} + \mathcal{R}_{\mathrm{des}}\right)\left(k^*_{\mathrm{p}} + k^*_{\mathrm{PDB}}\right)^{\eta-2}}{\left(k^*_{\mathrm{p}} + k^*_{\mathrm{PDB}} + k^*_{\mathrm{fm}} + k^*_{\mathrm{t}}c^{\mathrm{l}}_{\lambda_0} + \mathcal{R}_{\mathrm{e}}\right)^{\eta-1}}. \tag{D.12}$$

The average molecular weight of a monomer unit $M_{\mathrm{w,unit}}$, which is required for the calculation of the nucleation radius r_{nuc} and the growth rate of the particle $v(r)$, is determined from the reacted moles of monomers $n_{\mathrm{M}_{\mathrm{reac}i}}$ and the respective molecular weight M_{M_i}

$$M_{\mathrm{w,unit}} = \frac{\Sigma_{i=1}^{2} M_{\mathrm{M}_i} n_{\mathrm{M}_{\mathrm{reac}i}}}{\Sigma_{i=1}^{2} n_{\mathrm{M}_{\mathrm{reac}i}}}, \tag{D.13}$$

$$\frac{\mathrm{d}n_{\mathrm{M}_{\mathrm{reac}i}}}{\mathrm{d}t} = -\frac{\mathrm{d}n_{\mathrm{M}_i}}{\mathrm{d}t}. \tag{D.14}$$

D.2 Predicted particle radius in comparison to *in situ* DLS

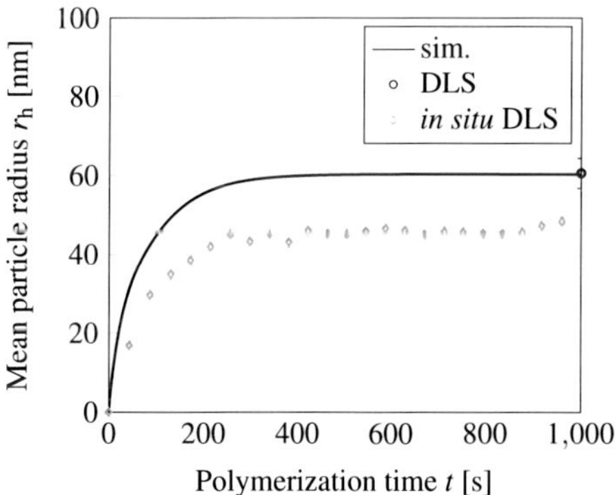

Figure D.1: Prediction of growth of average particle radius over polymerization time in comparison to experimental data from *in situ* DLS.

Figure D.1 shows the predicted particle radius over polymerization time in comparison to experimental data from DLS and *in situ* DLS. The predicted particle radius at the end of the simulation ($t = 1000\,\mathrm{s}$) satisfies the particle radius from DLS within the accuracy of the standard deviation. *In situ* DLS measurements are illustrated only for the sake of qualitative comparison. *In situ* DLS measurements require an interruption of stirring of the reaction mixture which might affect the particle growth. Further, high polymer concentration of the reaction mixture as well as turbidity affect the accuracy of *in situ* DLS measurements. Also, measurements of DLS and *in situ* DLS are measured with different angles (DLS: 90°, *in situ* DLS: 180°), which can cause deviations in the determined particle radius. Hence, these measurements are only used for qualitative comparison to the predicted particle growth and are not incorporated in the parameter estimation. The predicted average particle radius shows, besides a general offset, a similar trend as the particle growth from *in situ* DLS with regard to polymerization time.

D.3 Estimated parameter values and remarks

Table D.1: Estimated parameter values and additional remarks for evaluation.

Parameter	Parameter value	Comments
f_{e}	0.534	Estimated value is high compared to values from Nomura et al. (2005) ($f_{\mathrm{e}} = 10^{-4} - 10^{-3}$). This can be explained by the inaccuracies of pseudo-bulk models for homogeneous nucleation. In Eq. (3.13), the high $\mathcal{R}_{\mathrm{e}}^{\mathrm{p}}$ compensates $\mathcal{R}_{\mathrm{t}}^{\mathrm{p}}$, which would otherwise bring the polymerization to a hold right after initiation. Also, the high $\mathcal{R}_{\mathrm{e}}^{\mathrm{p}}$ causes radicals to enter a particle before they grow up to the critical chain length and precipitate, limiting precipitation to a short interval in the beginning (cf. Figure 3.3).
D_{P}	$10^{-16.65}\,\mathrm{m^2 s^{-1}}$	Low value indicates that the contribution of radical desorption is small.
$k_{\mathrm{t,diff11}}^{\mathrm{g}}$ $k_{\mathrm{t,diff22}}^{\mathrm{g}}$	$10^{-0.9}\,\mathrm{m^3 (mol\ s)^{-1}}$ $10^{1.64}\,\mathrm{m^3 (mol\ s)^{-1}}$	Same trend ($k_{\mathrm{t,diff11}}^{\mathrm{g}} < k_{\mathrm{t,diff22}}^{\mathrm{g}}$) as determined with two-phase model, but $k_{\mathrm{t,diff11}}^{\mathrm{g}}$ is higher than previously estimated due to ongoing radical entry.
r_{stable}	26.27 nm	Significantly higher than r_{nuc}, suggesting that particle aggregation is not limited to precursor particles and, despite late aggregation, a homogeneous PSD can be obtained.
W	$10^{4.44}$	Value appears to be in an average order of magnitude, given that it comprises a wide range of particle radii. Lower than values provided by Vale & McKenna (2009) for emulsion polymerization (10^5-10^6). Note that simulation results were not improved by a distinction among unstable/unstable and unstable/stable coagulation and in consequence, it was neglected. Further, parameter estimates of r_{stable} and W are highly correlated. Also, when coagulation is excluded ($\beta = 0$), no parameter values could be obtained by parameter estimation to achieve the final particle radius.

D.4 Variation of coagulation parameter values

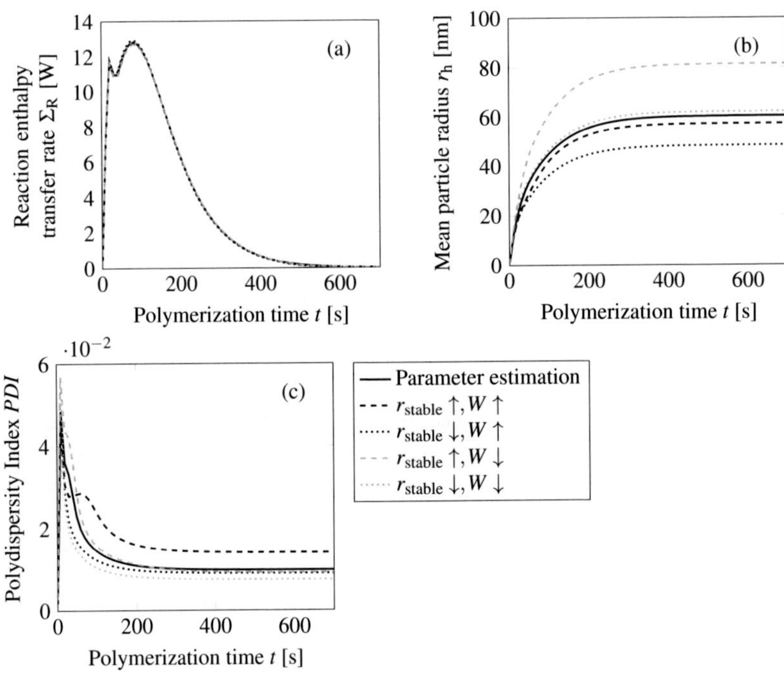

Figure D.2: Simulation study for variations of coagulation parameter values r_{stable} and W to illustrate their impact on Σ_R (a), r_h (b), and PDI (c).

Figure D.2 shows the results of a simulation study for different parameter variations for r_{stable} and W and their impact on the reaction enthalpy transfer rate Σ_R, the average microgel radius r_h and the polydispersity index PDI. Parameter estimation values were each increased as well as decreased, respectively ($r_{stable} = 26.27 \pm 10\,\text{nm}$, $W = 10^{4.44\pm0.5}$).
Variation of the parameter values has a negligible effect on the Σ_R. Increase of r_{stable} and decrease of W both facilitate particle aggregation and therefore result in larger particle radii (a). Increasing or decreasing both simultaneously does not affect the particle radius significantly as both influences eliminate each other. An increase of r_{stable} leads to a broader PDI, especially at the earlier stages of the process. Measurements of the particle size distribution over

the progress of the polymerization would improve the identifiability of the parameter values and indicate the order of magnitude of r_{stable} and W.

Appendix E

Hybrid KMC Model

E.1 Monomer and radical balances

The mass balances for the monomers are calculated by

$$\frac{\mathrm{d}n_{\mathrm{M}_i}}{\mathrm{d}t} = -V^1 \left(2 f_{\mathrm{I}}^1 k_{\mathrm{d}} c_{\mathrm{I}}^1 f_{\mathrm{M}_i}^1 + \sum_{j=1}^{2} \left(k_{\mathrm{p}ji}^1 + k_{\mathrm{fm}ji}^1 \right) c_{\lambda_0^j}^1 c_{\mathrm{M}_i}^1 \right) - \dot{n}_{\mathrm{M}_i}. \tag{E.1}$$

The concentrations of monomers in the liquid and particle phase are calculated from Eqs. (B.5), (B.6), and (B.7).

The balances for the radical in the liquid phase are formulated according to the respective radical species i and length $n = 1, ..., \eta - 1$ (cf. Eq. (B.8)):

$$\begin{aligned} \frac{\mathrm{d}n_{\mathrm{R}_n^i}^1}{\mathrm{d}t} = V^1 \Bigg(& \left(2 f_{\mathrm{I}}^l k_{\mathrm{d}} c_{\mathrm{I}}^1 f_{\mathrm{M}_i}^1 + \sum_{j=1}^{2} k_{\mathrm{fm}ji}^1 c_{\lambda_0^j}^1 c_{\mathrm{M}_i}^1 \right) \delta_{1,n} + \sum_{j=1}^{2} k_{\mathrm{p}ji}^1 c_{\mathrm{R}_{(n-1)}^j}^1 c_{\mathrm{M}_i}^1 \\ & - \sum_{j=1}^{2} (k_{\mathrm{p}ij}^1 + k_{\mathrm{fm}ij}^1) c_{\mathrm{R}_n^i}^1 c_{\mathrm{M}_j}^1 - \sum_{j=1}^{2} k_{\mathrm{td}ij}^1 c_{\mathrm{R}_n^i}^1 c_{\lambda_0^j}^1 \Bigg) \end{aligned} \tag{E.2}$$

The balances for precipitating radicals (R_η^i) and dead polymer (μ_0, μ_1) are used as formulated in Eqs. (B.9) and (B.11). The pendant double bonds attached to precipitating radicals are calculated according to Eqs. (B.12) and (B.13). The composition of the precipitating radicals is approximated by the reacted monomers $\mathrm{M}_{\mathrm{reac}}$ by recent propagation and chain transfer reactions,

$$\frac{\mathrm{d}n_{\mathrm{M}_{\mathrm{reac}i}}}{\mathrm{d}t} = V^1 \left(\sum_{j=1}^{2} \left(k_{\mathrm{p}ji}^1 + k_{\mathrm{fm}ji}^1 \right) c_{\lambda_0^j}^1 c_{\mathrm{M}_i}^1 \right). \tag{E.3}$$

E.2 Simulation results: Fitted hybrid KMC model

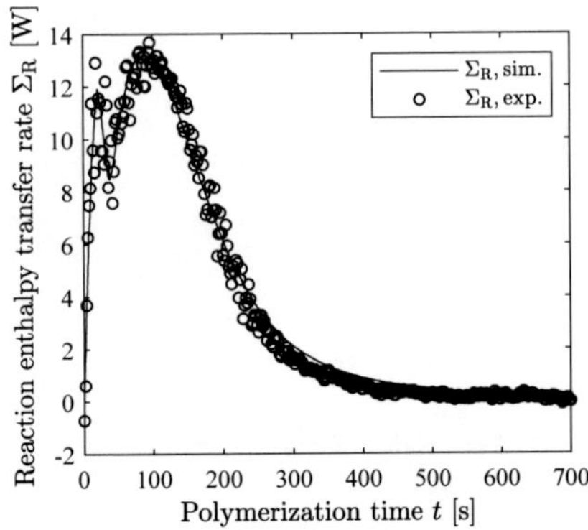

Figure E.1: Reaction enthalpy transfer rate for hybrid KMC model with adjusted parameter values. The parameter values for the diffusion limitation in the liquid phase were adjusted ($k^{g*}_{t,diff\,ii} = 0.5\; k^{g}_{t,diff\,ii}$), $k^{g}_{t,diff\,ii}$ from fitted terpolymerization precipitation model, cf. Table A.6

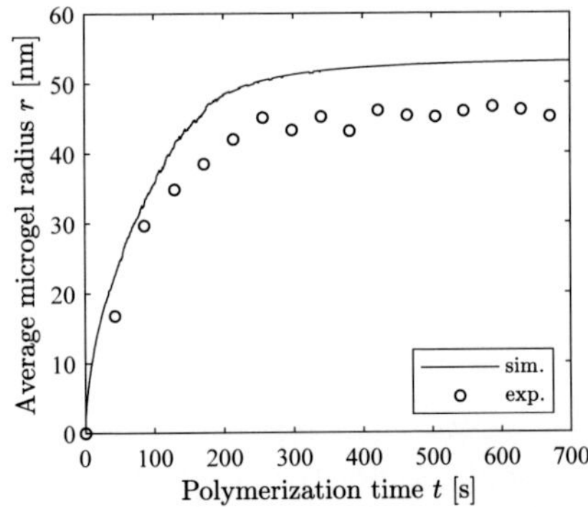

Figure E.2: Average particle radius predicted with hybrid KMC model with adjusted parameter values.

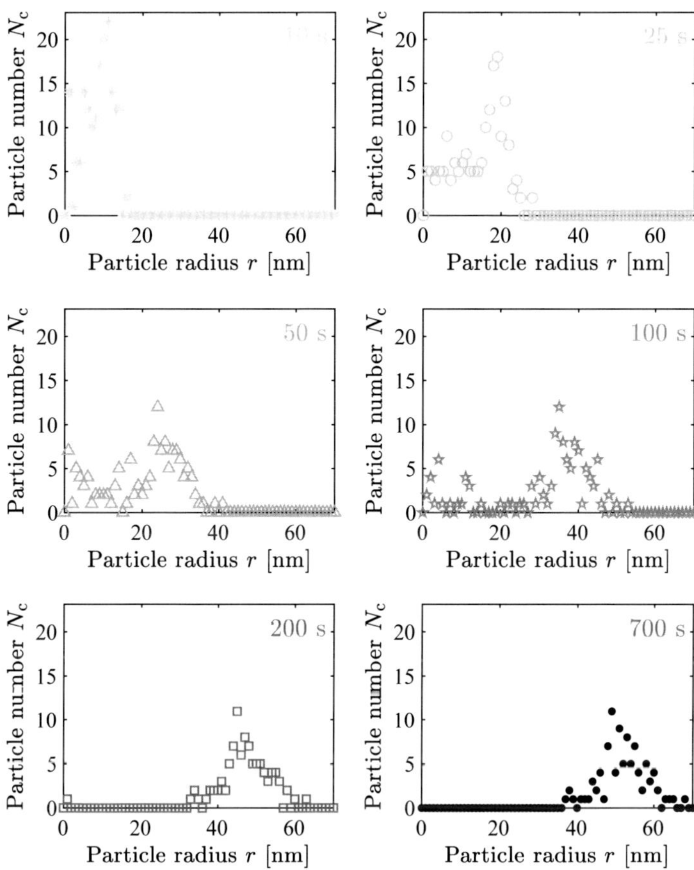

Figure E.3: Particle size distribution predicted with hybrid KMC model with adjusted parameter values.

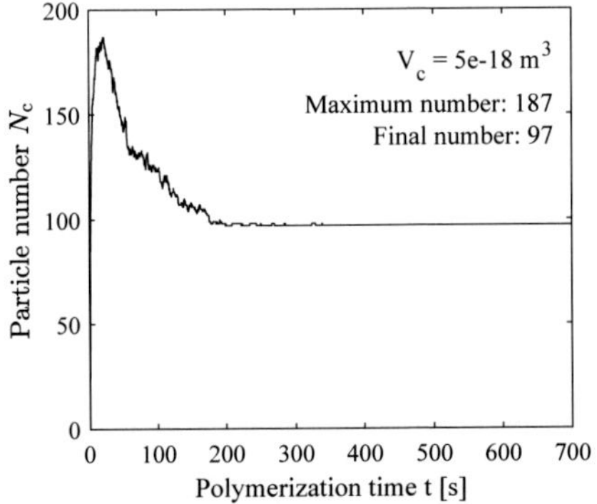

Figure E.4: Number of particles in the control volume V_c over polymerization time, *after* aggregation calculation (meso-scale simulation). Within the first 10 s, the particle number reaches a maximum of 187 particles. Then, the number of particles declines until approx. 200 s. After approx. 200 s, the number of particles remains constant. At the end of the simulation time, 97 particles remain in the control volume.

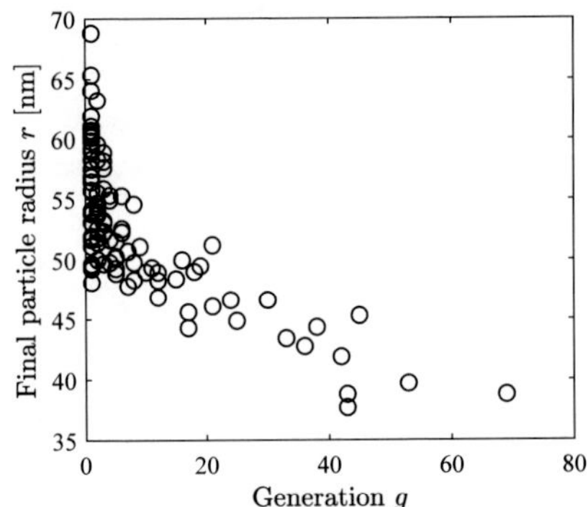

Figure E.5: Number of particles in the respective generation as predicted with the adjusted parameter values. The majority of particles of later generations was absorbed by particles formed in early generations. At the end of the simulation, mostly particles precipitated within the first 20 generations (5 s) remain.

Bibliography

Acciaro, R., Gilányi, T., & Varga, I. (2011). Preparation of monodisperse poly(*N*-isopropylacrylamide) microgel particles with homogenous cross-link density distribution. *Langmuir*, *27*, 7917–7925.

Agrawal, G., & Agrawal, R. (2018). Functional microgels: Recent advances in their biomedical applications. *Small*, *14*, 1801724.

Alfrey, T., & Goldfinger, G. (1944). The mechanism of copolymerization. *Journal of Chemical Physics*, *12*, 205–209.

Araújo, P. H. H., de la Cal, José Carlos, Asua, J. M., & Pinto, J. C. (2001). Modeling particle size distribution (PSD) in emulsion copolymerization reactions in a continuous loop reactor. *Macromolecular Theory and Simulations*, *10*, 769–779.

Arndt, M. C., & Sadowski, G. (2012). Modeling poly(*N*-isopropylacrylamide) hydrogels in water/alcohol mixtures with PC-SAFT. *Macromolecules*, *45*, 6686–6696.

Arosio, P., Mosconi, M., Storti, G., Banaszak, B., Hungenberg, K.-D., & Morbidelli, M. (2011a). Precipitation copolymerization of vinyl-imidazole and vinyl-pyrrolidone, 2 – Kinetic model. *Macromolecular Reaction Engineering*, *5*, 501–517.

Arosio, P., Mosconi, M., Storti, G., & Morbidelli, M. (2011b). Precipitation copolymerization of vinyl-imidazole and vinyl-pyrrolidone, 1 – Experimental analysis. *Macromolecular Reaction Engineering*, *5*, 490–500.

Avela, A., Poersch, H.-G., & Reichert, K.-H. (1990). Modelling the kinetics of the precipitation polymerization of acrylic acid. *Angewandte Makromolekulare Chemie*, *175*, 107–116.

Balaceanu, A., Demco, D. E., Möller, M., & Pich, A. (2011). Microgel heterogeneous morphology reflected in temperature-induced volume transition and ^{1}H high-resolution transverse relaxation NMR. The Case of Poly(*N*-vinylcaprolactam) Microgel. *Macromolecules*, *44*, 2161–2169.

Balaceanu, A., Mayorga, V., Lin, W., Schürings, M.-P., Demco, D. E., Böker, A., Winnik, M. A., & Pich, A. (2013). Copolymer microgels by precipitation polymerisation of *N*-vinylcaprolactam and *N*-isopropylacrylamides in aqueous medium. *Colloid and Polymer Science*, *291*, 21–31.

Brunier, B., Sheibat-Othman, N., Othman, S., Chevalier, Y., & Bourgeat-Lami, E. (2017). Modelling particle growth under saturated and starved conditions in emulsion polymerization. *The Canadian Journal of Chemical Engineering*, *95*, 208–221.

Buback, M., Gilbert, R. G., Russell, G. T., Hill, D. J. T., Moad, G., O'Driscoll, K. F., Shen, J., & Winnik, M. A. (1992). Consistent values of rate parameters in free radical polymerization systems. II. Outstanding dilemmas and recommendations. *Journal of Polymer Science Part A: Polymer Chemistry*, *30*, 851–863.

Bunyakan, C., Armanet, L., & Hunkeler, D. (1999). Precipitation polymerization of acrylic acid in toluene. II: Mechanism and kinetic modeling. *Polymer*, *40*, 6225–6234.

Chaffey-Millar, H., Stewart, D., Chakravarty, M. M. T., Keller, G., & Barner-Kowollik, C. (2007). A parallelised high performance monte carlo simulation approach for complex polymerisation kinetics. *Macromolecular Theory and Simulations*, *16*, 575–592.

Coen, E. M., Gilbert, R. G., Morrison, B. R., Leube, H., & Peach, S. (1998). Modelling particle size distributions and secondary particle formation in emulsion polymerisation. *Polymer*, *39*, 7099–7112.

Coen, E. M., Peach, S., Morrison, B. R., & Gilbert, R. G. (2004). First-principles calculation of particle formation in emulsion polymerization: Pseudo-bulk systems. *Polymer*, *45*, 3595–3608.

Cortez-Lemus, N. A., & Licea-Claverie, A. (2016). Poly(*N*-vinylcaprolactam), a comprehensive review on a thermoresponsive polymer becoming popular. *Progress in Polymer Science*, *53*, 1–51.

Costa, D., Valente, A. J., Miguel, M. G., & Queiroz, J. (2014). Plasmid DNA microgels for drug/gene co-delivery: A promising approach for cancer therapy. *Colloids and Surfaces A: Physicochemical and Engineering Aspects*, *442*, 181 – 190. Selected papers from the 26th European Colloid and Interface Society conference (26th ECIS 2012).

Costa, L. I., Storti, G., Morbidelli, M., Ferro, L., Galia, A., Scialdone, O., & Filardo, G. (2012). Copolymerization of VDF and HFP in supercritical carbon dioxide: A robust approach for modeling precipitation and dispersion kinetics. *Macromolecular Reaction Engineering*, *6*, 24–44.

Cramer, C. J., & Truhlar, D. G. (2008). A universal approach to solvation modeling. *Acc. Chem. Res.*, *41*, 760–768.

Crowley, T. J., Meadows, E. S., Kostoulas, E., & Doyle III, F. J. (2000). Control of particle size distribution described by a population balance model of semibatch emulsion polymerization. *Journal of Process Control*, *10*, 419–432.

Drache, M., Brandl, K., Reinhardt, R., & Beuermann, S. (2018). Ab initio kinetic Monte Carlo simulation of seeded emulsion polymerizations of styrene. *Physical chemistry chemical physics : PCCP*, .

Dubé, M. A., Soares, J. B. P., Penlidis, A., & Hamielec, A. E. (1997). Mathematical modeling of multicomponent chain-growth polymerizations in batch, semibatch, and continuous reactors: A review. *Industrial & Engineering Chemistry Research*, *36*, 966–1015.

Duracher, D., Elaïssari, A., & Pichot, C. (1999). Characterization of cross-linked poly(*N*-isopropylmethacrylamide) microgel latexes. *Colloid & Polymer Science*, *277*, 905–913.

Elizalde, O., Azpeitia, M., Reis, M. M., Asua, J. M., & Leiza, J. R. (2005). Monitoring emulsion polymerization reactors: Calorimetry versus Raman spectroscopy. *Industrial & Engineering Chemistry Research*, *44*, 7200–7207.

Enright, T., & Zhu, S. (2000). Modeling and semi-batch control of cross-link density distribution in the free-radical copolymerization of vinyl/divinyl monomers. *Macromolecular Theory and Simulations*, *9*, 196–206.

Fernandez-Nieves, A., Wyss, H., Mattsson, J., & Weitz, D. A. (2011). Microgel suspensions: Fundamentals and applications. John Wiley & Sons.

Fuxman, A. M., McAuley, K. B., & Schreiner, L. J. (2003). Modeling of free-radical crosslinking copolymerization of acrylamide and N,N'-methylenebis(acrylamide) for radiation dosimetry. *Macromolecular Theory and Simulations*, *12*, 647–662.

Gao, J., & Frisken, B. J. (2003). Influence of reaction conditions on the synthesis of self-cross-linked *N*-isopropylacrylamide microgels. *Langmuir*, *19*, 5217–5222.

Gavrilov, A. A., Rudyak, V. Y., & Chertovich, A. V. (2020). Computer simulation of the core-shell microgels synthesis via precipitation polymerization. *Journal of Colloid and Interface Science*, *574*, 393 – 398.

Gervasio, M., & Lu, K. (2019). Monte carlo simulation modeling of nanoparticle–polymer cosuspensions. *Langmuir*, *35*, 161–170.

Gillespie, D. T. (1976). A general method for numerically simulating the stochastic time evolution of coupled chemical reactions. *Journal of Computational Physics*, *22*, 403–434.

Green, D. W., & Perry, R. H. (2008). Perrys's chemical engineers' handbook. (8th ed.). New York: McGraw-Hill.

Hamzehlou, S., Ballard, N., Carretero, P., Paulis, M., Asua, J. M., Reyes, Y., & Leiza, J. R. (2014a). Mechanistic investigation of the simultaneous addition and free-radical polymerization in batch miniemulsion droplets: Monte Carlo simulation versus experimental data in polyurethane/acrylic systems. *Polymer*, *55*, 4801 – 4811.

Hamzehlou, S., Reyes, Y., & Leiza, J. R. (2012). Detailed microstructure investigation of acrylate/methacrylate functional copolymers by kinetic monte carlo simulation. *Macromolecular Reaction Engineering*, *6*, 319–329.

Hamzehlou, S., Reyes, Y., & Leiza, J. R. (2013). A new insight into the formation of polymer networks: A Kinetic Monte Carlo simulation of the cross-Linking polymerization of S/DVB. *Macromolecules*, *46*, 9064–9073.

Hamzehlou, S., Reyes, Y., & Leiza, J. R. (2014b). Modeling the mini-emulsion copolymerization of *N*-butyl acrylate with a water-soluble monomer: A Monte Carlo approach. *Industrial & Engineering Chemistry Research*, *53*, 8996–9003.

Hansen, F. K., & Ugelstad, J. (1978). Particle nucleation in emulsion polymerization. I. A theory for homogeneous nucleation. *Journal of Polymer Science: Polymer Chemistry Edition*, *16*, 1953–1979.

Harada, M., Nomura, M., Eguchi, W., & Nagata, S. (1971). Studies of the effect of polymer particles on emulsion polymerization. *Journal of Chemical Engineering of Japan*, *4*, 54–60.

Hoare, T., & McLean, D. (2006a). Kinetic prediction of functional group distributions in thermosensitive microgels. *The Journal of Physical Chemistry B*, *110*, 20327–20336.

Hoare, T., & McLean, D. (2006b). Multi-component kinetic modeling for controlling local compositions in thermosensitive polymers. *Macromolecular Theory and Simulations*, *15*, 619–632.

Imaz, A., & Forcada, J. (2008a). *N*-vinylcaprolactam-based microgels: Effect of the concentration and type of cross-linker. *Journal of Polymer Science Part A: Polymer Chemistry*, *46*, 2766–2775.

Imaz, A., & Forcada, J. (2008b). *N*-vinylcaprolactam-based microgels: Synthesis and characterization. *Journal of Polymer Science Part A: Polymer Chemistry*, *46*, 2510–2524.

Imaz, A., & Forcada, J. (2010). *N*-vinylcaprolactam-based microgels for biomedical applications. *Journal of Polymer Science Part A: Polymer Chemistry*, *48*, 1173–1181.

Immanuel, C. D., Cordeiro, C. F., Sundaram, S. S., Meadows, E. S., Crowley, T. J., & Doyle, F. J. (2002). Modeling of particle size distribution in emulsion co-polymerization: comparison with experimental data and parametric sensitivity studies. *Computers & Chemical Engineering*, *26*, 1133–1152.

Immanuel, C. D., Doyle, F. J., Cordeiro, C. F., & Sundaram, S. S. (2003). Population balance PSD model for emulsion polymerization with steric stabilizers. *AIChE Journal*, *49*, 1392–1404.

Immanuel, C. D., & Doyle III, F. J. (2003). Computationally efficient solution of population balance models incorporating nucleation, growth and coagulation: Application to emulsion polymerization. *Chemical Engineering Science*, *58*, 3681–3698.

Janssen, F. A. L., Kather, M., Kröger, L. C., Mhamdi, A., Leonhard, K., Pich, A., & Mitsos, A. (2017). Synthesis of poly(*N*-vinylcaprolactam)-based microgels by precipitation polymerization: Process modeling and experimental validation. *Industrial & Engineering Chemistry Research*, *56*, 14545–14556.

Janssen, F. A. L., Kather, M., Ksiazkiewicz, A., Pich, A., & Mitsos, A. (2019). Synthesis of poly(*N*-vinylcaprolactam)-based microgels by precipitation polymerization: Pseudo-bulk model for particle growth and size distribution. *ACS Omega*, *4*, 13795–13807.

Janssen, F. A. L., Ksiazkiewicz, A., Kather, M., Kröger, L. C., Mhamdi, A., Leonhard, K., Pich, A., & Mitsos, A. (2018). Kinetic modeling of precipitation terpolymerization for functional microgels. In *28th European Symposium on Computer Aided Process Engineering* (pp. 109–114). Elsevier Volume 43 of *Computer Aided Chemical Engineering*.

Joback, K. G., & Reid, R. C. (1987). Estimation of pure-component properties from group-contributions. *Chemical Engineering Communications*, *57*, 233–243.

Jung, F., Janssen, F. A. L., Ksiazkiewicz, A., Caspari, A., Mhamdi, A., Pich, A., & Mitsos, A. (2019a). Identifiability analysis and parameter estimation of microgel synthesis: A set-membership approach. *Industrial & Engineering Chemistry Research*, *58*, 13675–13685.

Jung, F., Ksiazkiewicz, A., Mhamdi, A., Pich, A., & Mitsos, A. (2019b). Model-based prediction of the hydrodynamic radius of collapsed microgels and experimental validation. *Chemical Engineering Journal*, *378*, 121740.

Karg, M., Pich, A., Hellweg, T., Hoare, T., Lyon, L. A., Crassous, J. J., Suzuki, D., Gumerov, R. A., Schneider, S., Potemkin, I. I., & Richtering, W. (2019). Nanogels and microgels: From model colloids to applications, recent developments, and future trends. *Langmuir*, *35*, 6231–6255.

Khalili, S., Lin, Y., Armaou, A., & Matsoukas, T. (2010). Constant number monte carlo simulation of population balances with multiple growth mechanisms. *AIChE Journal*, *56*, 3137–3145.

Kiparissides, C. (1996). Polymerization reactor modeling: A review of recent developments and future directions. *Chemical Engineering Science*, *51*, 1637–1659.

Kiparissides, C. (2006). Challenges in particulate polymerization reactor modeling and optimization: A population balance perspective. *Journal of Process Control*, *16*, 205–224.

Kiparissides, C., Achilias, D. S., & Frantzikinakis, C. E. (2002). The effect of oxygen on the kinetics and particle size distribution in vinyl chloride emulsion polymerization. *Industrial & Engineering Chemistry Research*, *41*, 3097–3109.

Klamt, A. (1995). Conductor-like screening model for real solvents: A new approach to the quantitative calculation of solvation phenomena. *The Journal of Physical Chemistry*, *99*, 2224–2235.

Kremer, K., & Binder, K. (1988). Monte carlo simulation of lattice models for macromolecules. *Computer Physics Reports*, *7*, 259–310.

Kröger, L. C., Kopp, W. A., & Leonhard, K. (2017). Prediction of chain propagation rate constants of polymerization reactions in aqueous NIPAM/BIS and VCL/BIS systems. *The Journal of Physical Chemistry B*, *121*, 2887–2895.

Ksiazkiewicz, A. N., Bering, L., Jung, F., Wolter, N. A., Viell, J., Mitsos, A., & Pich, A. (2020). Closing the 1-5 μm gap: Temperature-programmed, fed-batch synthesis of μm-sized microgels. *Chemical Engineering Journal*, *379*, 122293.

Kwon, S., Bae, W., & Kim, H. (2005). High-pressure phase behavior of CO_2 + N-vinyl caprolactam and CO_2 + N-methyl caprolactam systems. *Journal of Chemical & Engineering Data*, *50*, 1560–1563.

Lattuada, M., Sandkühler, P., Wu, H., Sefcik, J., & Morbidelli, M. (2004). Kinetic modeling of aggregation and gel formation in quiescent dispersions of polymer colloids. *Macromolecular Symposia*, *206*, 307–320.

Lazzari, S., Pfister, D., Diederich, V., Kern, A., & Storti, G. (2014). Modeling of acrylamide/*N*,*N*′-methylenebisacrylamide solution copolymerization. *Industrial & Engineering Chemistry Research*, *53*, 9035–9048.

Lemos, T., Melo, P. A., & Pinto, J. C. (2015). Stochastic modeling of polymer microstructure from residence time distribution. *Macromolecular Reaction Engineering*, *9*, 259–270.

Lohaus, T., de Wit, P., Kather, M., Menne, D., Benes, N. E., Pich, A., & Wessling, M. (2017). Tunable permeability and selectivity: Heatable inorganic porous hollow fiber membrane with a thermo-responsive microgel coating. *Journal of Membrane Science*, *539*, 451–457.

Loschen, C., & Klamt, A. (2014). Prediction of solubilities and partition coefficients in polymers using COSMO-RS. *Industrial & Engineering Chemistry Research*, *53*, 11478–11487.

Maldonado-Parra, F. D. (2019). Microscale modeling of the growth of microgels by precipitation polymerization. Dissertation Rheinisch-Westfälische Technische Hochschule Aachen Aachen.

Marien, Y. W., Van Steenberge, P. H. M., Pich, A., & D'hooge, D. R. (2019a). Coupled stochastic simulation of the chain length and particle size distribution in miniemulsion radical copolymerization of styrene and *N*-vinylcaprolactam. *Reaction Chemistry & Engineering*, *4*, 1935–1947.

Marien, Y. W., Van Steenberge, P. H. M., R. D'hooge, D., & Marin, G. B. (2019b). Particle by particle kinetic monte carlo tracking of reaction and mass transfer events in miniemulsion free radical polymerization. *Macromolecules*, *52*, 1408–1423.

Mayo, F. R., & Lewis, F. M. (1944). Copolymerization. I. A basis for comparing the behavior of monomers in copolymerization; The copolymerization of styrene and methyl methacrylate. *Journal of the American Chemical Society*, *66*, 1594–1601.

Meimaroglou, D., & Kiparissides, C. (2014). Review of Monte Carlo methods for the prediction of distributed molecular and morphological polymer properties. *Industrial & Engineering Chemistry Research*, *53*, 8963–8979.

Menne, D., Pitsch, F., Wong, J. E., Pich, A., & Wessling, M. (2014). Temperature-modulated water filtration using microgel-functionalized hollow-fiber membranes. *Angewandte Chemie International Edition*, *53*, 5706–5710.

Meurer, R. A., Kemper, S., Knopp, S., Eichert, T., Jakob, F., Goldbach, H. E., Schwaneberg, U., & Pich, A. (2017). Biofunctional microgel-based fertilizers for controlled foliar delivery of nutrients to plants. *Angewandte Chemie International Edition*, *56*, 7380–7386.

Meyer-Kirschner, J., Kather, M., Ksiazkiewicz, A., Pich, A., Mitsos, A., & Viell, J. (2018). Monitoring microgel synthesis by copolymerization of *N*-isopropylacrylamide and *N*-vinylcaprolactam via in-line Raman spectroscopy and Indirect Hard Modeling. *Macromolecular Reaction Engineering*, *80*, 1700067.

Meyer-Kirschner, J., Kather, M., Pich, A., Engel, D., Marquardt, W., Viell, J., & Mitsos, A. (2016). In-line monitoring of monomer and polymer content during microgel synthesis using precipitation polymerization via Raman spectroscopy and Indirect Hard Modeling. *Applied Spectroscopy*, *70*, 416–426.

Mueller, P., Storti, G., & Morbidelli, M. (2005). The reaction locus in supercritical carbon dioxide dispersion polymerization. The case of poly(methyl methacrylate). *Chemical Engineering Science*, *60*, 377–397.

Nasresfahani, A., & Hutchinson, R. A. (2018). Modeling the distribution of functional groups in semibatch radical copolymerization: An accelerated stochastic approach. *Industrial & Engineering Chemistry Research*, *57*, 9407–9419.

Nomura, M., Tobita, H., & Suzuki, K. (2005). Emulsion polymerization: Kinetic and mechanistic aspects. In M. Okubo (Ed.), *Polymer Particles* (pp. 1–128). Berlin, Heidelberg: Springer Berlin Heidelberg Volume 175 of *Advances in Polymer Science*.

Odian, G. G. (2004). Principles of polymerization. (Fourth edition ed.). Hoboken, N.J.: Wiley.

Pelton, R. (2000). Temperature-sensitive aqueous microgels. *Advances in Colloid and Interface Science*, *85*, 1–33.

Pich, A., & Richtering, W. (2011). Microgels by precipitation polymerization: Synthesis, characterization, and functionalization. In A. Pich, & W. Richtering (Eds.), *Chemical Design of Responsive Microgels* (pp. 1–37). Berlin, Heidelberg: Springer Berlin Heidelberg Volume 234 of *Advances in Polymer Science*.

Pich, A., Tessier, A., Boyko, V., Lu, Y., & Adler, H.-J. P. (2006). Synthesis and characterization of poly(vinylcaprolactam)-based microgels exhibiting temperature and pH-sensitive properties. *Macromolecules*, *39*, 7701–7707.

Plamper, F. A., & Richtering, W. (2017). Functional microgels and microgel systems. *Accounts of chemical research*, *50*, 131–140.

Platkowski, K., & Reichert, K.-H. (1999). Application of monte carlo methods for modelling of polymerization reactions. *Polymer*, *40*, 1057–1066.

Ramos, J., Imaz, A., & Forcada, J. (2012). Temperature-sensitive nanogels: Poly(*N*-vinylcaprolactam) versus poly(*N*-isopropylacrylamide). *Polymer Chemistry*, *3*, 852–856.

Sajjadi, S. (2003). Particle formation under monomer-starved conditions in the semibatch emulsion polymerisation of styrene. Part II. Mathematical modelling. *Polymer*, *44*, 223–237.

Sajjadi, S. (2009). Population balance modeling of particle size distribution in monomer-starved semibatch emulsion polymerization. *AIChE Journal*, *55*, 3191–3205.

Saunders, B. R., Laajam, N., Daly, E., Teow, S., Hu, X., & Stepto, R. (2009). Microgels: From responsive polymer colloids to biomaterials. *Advances in Colloid and Interface Science*, *147-148*, 251–262.

Saunders, B. R., & Vincent, B. (1999). Microgel particles as model colloids: theory, properties and applications. *Advances in Colloid and Interface Science*, *80*, 1–25.

Schneider, F., Balaceanu, A., Feoktystov, A., Pipich, V., Wu, Y., Allgaier, J., Pyckhout-Hintzen, W., Pich, A., & Schneider, G. J. (2014). Monitoring the internal structure of poly(*N*-vinylcaprolactam) microgels with variable cross-link concentration. *Langmuir*, *30*, 15317–15326.

Sheibat-Othman, N., Vale, H. M., Pohn, J. M., & McKenna, T. F. L. (2017). Is modeling the PSD in emulsion polymerization a finished problem? An overview. *Macromolecular Reaction Engineering*, *11*, 1600059.

Smith, W. V., & Ewart, R. H. (1948). Kinetics of emulsion polymerization. *Journal of Chemical Physics*, *16*, 592–599.

Šomvársky, J., & Dušek, K. (1994). Kinetic monte-carlo simulation of network formation. *Polymer Bulletin*, *33*, 369–376.

Stieger, M., Richtering, W., Pedersen, J. S., & Lindner, P. (2004). Small-angle neutron scattering study of structural changes in temperature sensitive microgel colloids. *Journal of Chemical Physics*, *120*, 6197–6206.

Storti, G., Carrà, S., Morbidelli, M., & Vita, G. (1989). Kinetics of multimonomer emulsion polymerization. The pseudo-homopolymerization approach. *Journal of Applied Polymer Science*, *37*, 2443–2467.

Stubbs, J., Carrier, R., & Sundberg, D. C. (2008). Monte Carlo simulation of emulsion polymerization kinetics and the evolution of latex particle morphology and polymer chain architecture. *Macromolecular Theory and Simulations*, *17*, 147–162.

Thickett, S. C., & Gilbert, R. G. (2007). Emulsion polymerization: State of the art in kinetics and mechanisms. *Polymer*, *48*, 6965–6991.

Thorne, J. B., Vine, G. J., & Snowden, M. J. (2011). Microgel applications and commercial considerations. *Colloid and Polymer Science*, *289*, 625–646.

Tobita, H., & Hamielec, A. E. (1989). Modeling of network formation in free radical polymerization. *Macromolecules*, *22*, 3098–3105.

Tobita, H., & Yanase, F. (2007). Monte carlo simulation of controlled/living radical polymerization in emulsified systems. *Macromolecular Theory and Simulations*, *16*, 476–488.

Tripathi, A. K., Tsavalas, J. G., & Sundberg, D. C. (2014). Monte Carlo simulations of free radical polymerizations with divinyl cross-linker: Pre- and postgel simulations of reaction kinetics and molecular structure. *Macromolecules*, *48*, 184–197.

Uhelská, L., Chorvát, D., Hutchinson, R. A., Santanakrishnan, S., Buback, M., & Lacík, I. (2014). Radical propagation kinetics of *N*-vinylpyrrolidone in organic solvents studied by Pulsed-Laser Polymerization-Size-Exclusion Chromatography (PLP-SEC). *Macromolecular Chemistry and Physics*, *215*, 2327–2336.

Vale, H. M., & McKenna, T. F. (2005). Modeling particle size distribution in emulsion polymerization reactors. *Progress in Polymer Science*, *30*, 1019–1048.

Vale, H. M., & McKenna, T. F. (2009). Particle formation in vinyl chloride emulsion polymerization: Reaction modeling. *Industrial & Engineering Chemistry Research*, *48*, 5193–5210.

Van Steenberge, P., D'hooge, D., Reyniers, M.-F., & Marin, G. (2014). Improved kinetic monte carlo simulation of chemical composition-chain length distributions in polymerization processes. *Chemical Engineering Science*, *110*, 185–199.

Varga, I., Gilányi, T., Mészáros, R., Filipcsei, G., & Zrínyi, M. (2001). Effect of cross-link density on the internal structure of poly(*N*-isopropylacrylamide) microgels. *The Journal of Physical Chemistry B*, *105*, 9071–9076.

Virtanen, O. L. J., Ala-Mutka, H. M., & Richtering, W. (2015). Can the reaction mechanism of radical solution polymerization explain the microgel final particle volume in precipitation polymerization of *N*-isopropylacrylamide? *Macromolecular Chemistry and Physics*, *216*, 1431–1440.

Virtanen, O. L. J., Brugnoni, M., Kather, M., Pich, A., & Richtering, W. (2016). The next step in precipitation polymerization of *N*-isopropylacrylamide: Particle number density control by monochain globule surface charge modulation. *Polymer Chemistry*, *7*, 5123–5131.

Virtanen, O. L. J., Kather, M., Meyer-Kirschner, J., Melle, A., Radulescu, A., Viell, J., Mitsos, A., Pich, A., & Richtering, W. (2019). Direct monitoring of microgel formation during precipitation polymerization of *N*-isopropylacrylamide using in situ sans. *ACS Omega, 4*, 3690–3699.

Virtanen, O. L. J., & Richtering, W. (2014). Kinetics and particle size control in non-stirred precipitation polymerization of *N*-isopropylacrylamide. *Colloid and Polymer Science, 292*, 1743–1756.

Wang, L., & Broadbelt, L. J. (2011). Tracking explicit chain sequence in kinetic monte carlo simulations. *Macromolecular Theory and Simulations, 20*, 54–64.

Wang, R., Lin, T.-S., Johnson, J. A., & Olsen, B. D. (2017). Kinetic monte carlo simulation for quantification of the gel point of polymer networks. *ACS Macro Letters, 6*, 1414–1419.

Welsch, N., Ballauff, M., & Lu, Y. (2010). Microgels as nanoreactors: Applications in catalysis. (pp. 129–163). Volume 234.

Wieme, J., D'hooge, D. R., Reyniers, M.-F., & Marin, G. B. (2009). Importance of radical transfer in precipitation polymerization: The case of vinyl chloride suspension polymerization. *Macromolecular Reaction Engineering, 3*, 16–35.

Wilke, C. R., & Chang, P. (1955). Correlation of diffusion coefficients in dilute solutions. *AIChE Journal, 1*, 264–270.

Wolff, H. J. M., Kather, M., Breisig, H., Richtering, W., Pich, A., & Wessling, M. (2018). From batch to continuous precipitation polymerization of thermoresponsive microgels. *ACS Applied Materials & Interfaces, 10*, 24799–24806.

Wu, X., Pelton, R. H., Hamielec, A. E., Woods, D. R., & McPhee, W. (1994). The kinetics of poly(*N*-isopropylacrylamide) microgel latex formation. *Colloid & Polymer Science, 272*, 467–477.

Xie, T., & Hamielec, A. E. (1993). Modelling free-radical copolymerization kinetics – Evaluation of the pseudo-kinetic rate constant method, 1. Molecular weight calculations for linear copolymers. *Die Makromolekulare Chemie, Theory and Simulations, 2*, 421–454.

Zhu, S., Hamielec, A. E., & Pelton, R. H. (1993). Modelling of crosslinking and cyclization in free-radical copolymerization of vinyl/divinyl monomers. *Die Makromolekulare Chemie, Theory and Simulations, 2*, 587–604.